EcoLinking

EcoLinking

Everyone's Guide to Online Environmental Information

Don Rittner

PEACHPIT PRESS
BERKELEY, CALIFORNIA

*Dedicated
to
Nancy and Christopher*

ECOLINKING
Don Rittner

PEACHPIT PRESS, INC.
2414 Sixth St.
Berkeley, CA 94710
(510) 548-4393
(510) 548-5991 (fax)

Copyright © 1992 by Don Rittner
Cover design by Tony Lane
Cover illustration by Brooke Scudder
Interior design by Olav Martin Kvern
Satellite images © 1992 Rick Young

All rights reserved. No part of this book may be reproduced in any form or by any means, electronic, mechanical, photocopying, recording, or otherwise, without the prior written permission of the publisher. For information, contact Peachpit Press.

NOTICE OF LIABILITY:
The information in this book is distributed on an "As is" basis, without warranty. While every precaution has been taken in the preparation of this book, neither the author nor Peachpit Press, Inc., shall have any liability to any person or entity with respect to any liability, loss, or damage caused or alleged to be caused directly or indirectly by the instructions contained in this book or by the computer software and hardware products described herein.

TRADEMARKS:
Omnet and SCIENCEnet are registered in the US Patent & Trademark Office. Fido, FidoNet and the dog-with-diskette are US registered trademarks of Tom Jennings, Box 77731, San Francisco, CA 94107, USA. Throughout this book, other trademarked names are used. Rather than put a trademark symbol in every occurrence of a trademarked name, we are using the names only in an editorial fashion and to the benefit of the trademark owner, with no intention of infringement of the trademark.

ISBN 0-938151-35-5

0 9 8 7 6 5 4 3 2 1

Printed and bound in the United States of America

Printed on recycled paper

Contents

Preface vii

Acknowledgments xi

PART I Getting Online 1

   1 How to Use This Book 3
   2 The Basics 9

PART II Global Networks 27

   3 FidoNet 31
   4 BITNET 51
   5 Usenet 63
   6 Internet 79

PART III Electronic Bulletin Boards 97

   7 Electronic Bulletin Boards 99

PART IV Commercial Online Services 145

   8 America Online 147
   9 CompuServe 163
 10 EcoNet 177
 11 GEnie 201
 12 The WELL 205

PART V Libraries That Never Close 219

13 Online Research Databases 221

14 CD-ROM: Low-Cost Information Storage for
    Environmental Research 257

15 Instant Access to Environmental News 267

Appendix A Selected Communications Software 291

Appendix B Internet Mailing Lists 295

Appendix C Gateway Services to the Networks 315

Appendix D Sample BBS Session 319

Appendix E Recommended Reading 341

Index 343

We are going to have to find ways of organizing ourselves cooperatively, sanely, scientifically, harmonically and in regenerative spontaneity with the rest of humanity around earth.... We are not going to be able to operate our spaceship earth successfully nor for much longer unless we see it as a whole spaceship and our fate as common.

Buckminster Fuller

Preface

We are entering the nineties with an unprecedented ability to work with fellow environmentalists, scientists, and concerned citizens around the world by using personal computers.

This book will show you how!

We have painfully learned over the last fifty years that our actions have poisoned our water, our air, our land, and our people. We cannot repair this damage if we act, as organizations or individuals, in a vacuum. The extent of the damage dictates that we work together to get our global house in order. It is vitally important in the nineties to realize that we can think and act globally as well as locally.

But how?

Through "EcoLinking": the use of computer technology by scientists, environmentalists, and concerned citizens around the globe to share ideas and research on environmental issues. In the interest of pooling individual resources, this form of communication links concerned organizations and individuals both to each other and to the wealth of environmental and scientific information residing on personal computers, minicomputers, and mainframes everywhere.

The affordability of personal computers in recent years has spawned a surge of interest in online communication because of the power it brings to every computer user — power in the form of speed of communication, global reach, a forum for idea sharing, and a repository for information.

It's easy to recognize computer networking as a powerful vehicle for change. It's easy to imagine EcoLinking creating the opportunity for individuals and organizations to join forces and lobby their particular causes, reaching millions of people in a matter of minutes.

Online communication broadens your reach as an environmentally conscientious individual. Your chances of finding people with similar interests or with the right answers to your questions increase exponentially. Online communication instantly spans oceans and continents. It is not constrained by office hours, and it gives every participant equal access regardless of age, disability, race, creed, or sex. What matters is the meaning of the words and ideas that are transmitted across the digital frontier.

This book has two major purposes: to introduce you to the computer-based tools you need to become part of this global community, and to link you up with other environmentalists, scientists, and concerned citizens throughout the world.

You don't have to be a computer nerd or a technojunkie to take advantage of the power of today's communications technology. All you need is access to a personal computer, some inexpensive software, a modem, and your telephone.

Once you see how easy it is to get online, you will be able to join with environmentalists, scientists, and concerned citizens from Albany to Zambia in shaping the world's environmental agenda as we enter the twenty-first century.

This book is an action manual, a how-to book. It is long on practical matters and short on philosophy. It assumes that you know little about computers but want to learn more about online communications as a tool. The book provides the technical basics you need to get online with ease and cuts through the jargon of telecommunications whenever possible. You will soon learn how to travel around the planet as an electronic explorer, no longer confined by time or geographic boundaries.

It doesn't matter whether you are a scientist, an activist, a housewife, or a student committed to providing a cleaner and healthier world environment—the power of networking is open to all.

This book will arm you with an arsenal of facts and friends with which to wage war against the destruction of our planet and its resources. Computer-based technology has given you the power to accumulate knowledge from a diversity of sources and present your case to a global audience.

Never before has one individual been able to communicate so much to so many.

Don Rittner
Schenectady, New York

Acknowledgments

An effort of this magnitude would not be possible without the help of many people and organizations.

The following people all have contributed to the preparation of this book. My thanks to: Adam Angst, Anna Lange (DASNET), Bart Barton, Bill Earle (CIS); Bill Leland and Jill Small (Econet), Carol Reid, Charlotte Mooers (CSNET), David Dodell, David Hitt (Omnet), Donnalyn Frey (UUNET), Douglas Smith, Elizabeth Morley (Silverplatter), Gene Spafford, George Peace, Jean Polly, Joe Houghtaling, Judy Trimarchi, Jim Crabtree (ENS), K. Mulvey (KI), Karen Roubicek, Kathy Ryan, Larry Pina, Marianne Pina, Max Burgiss, Michael Stein (National Toxics Campaign), Michelle Hanson (OSHA); Mike Fuchs (Echolist), Nancy Benz, Patricia MacParland (NewsNet), Patricia Wagner (BRS), Paul Werner, Phil Eschallier (nixpub), Richard Baker (CompuServe), Robert Harper, Sarah Radick (Orbit), Sue Darbyshire, Tim Pozar, Tom Jennings, Tracy L. LaQuey, Victoria Wakefield (UPI).

Thanks to Wendy Monroe for researching and writing the case studies on pages 72, 186, 190, 222, 228, 271, and 280.

Special thanks go to Kathy Ryan, Larry & Marianne Pina, and Charlotte Mooers for their review, guidance, and editing.

Thanks to all the system operators of the dozens of bulletin boards featured in this book. Their long hard dedicated work is noted and appreciated.

I am especially grateful to my editor Kathy Ryan who believed in this project as much as I did, and to my publisher Ted Nace.

For those I inadvertently missed, please accept my apologies.

PART I

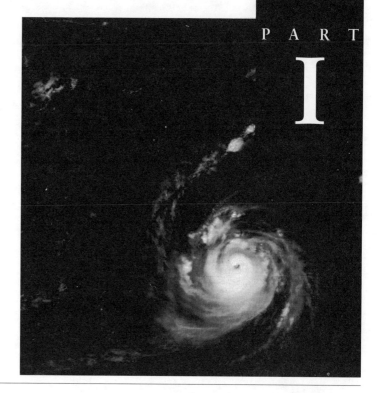

Getting Online

CHAPTER 1

> *A Land Ethic ... reflects a conviction of individual responsibility for the health of the land. Health is the capacity of the land for self-renewal. Conservation is our effort to understand and preserve this capacity.*
>
> Aldo Leopold,
> *A Sand County Almanac*, 1949

How to Use This Book

"Environmentalism" is the awareness that the social, economic, political, and cultural dynamics of humanity must work in harmony with the natural world, maintaining a delicate balance between what we take from and what we give back to our planet. This book will introduce you to the computer-based communication tools you need to conduct environmental research and networking more effectively. Using those tools, you will communicate with fellow environmentalists, scientists, and concerned citizens around the world. You will have access to the best and most current published and unpublished research in a variety of disciplines in minutes, no matter where it is. And ultimately, you will participate within a worldwide community in finding solutions to providing a sustainable global environment.

Computer networking is nothing new. It has been occurring for more than a decade on the expensive mainframe computers of corporations, academia, and government agencies. Thanks to the rise in popularity—and subsequent lowering in price—of personal computers, the power of computer networking is now available to the general public. The latest U.S. Census Bureau figures show that some 75 million Americans alone have access to a personal computer and that almost one-quarter of those have a modem.

Furthermore, much of the complexity of computer networking has been reduced so that only a basic knowledge of computing is required, resulting in a dramatic increase in the number of people using online resources as part of their everyday routines.

As an example, look at what occurs on a typical day in the networking community: Thousands of letters, reports, and business plans are sent electronically around the world, reaching their destinations in seconds. Environmental groups post requests for information and calls to action on computer bulletin boards for thousands to see. Scientists debate theories in live conferences online. A research technician asks for help solving a lab problem and gets answers from colleagues around the world. Activists form alliances with other groups separated geographically by hundreds of miles. Students conduct online searches through thousands of pages of published literature in seconds. Journalists and their editors send copy back and forth for editing and approval. A biologist downloads a needed software program from a bulletin board into his or her computer. People help other people in need through electronic mail.

All you need to become a part of this exciting global community are

- personal computer,
- a modem,
- communications software, and
- a telephone line.

For this book, I have selected just a few of the hundreds of electronic resources that are available for your use in conducting research, promoting activism, and finding solutions to the many environmental problems facing us today. All of these resources can provide you with the facts and sound arguments you need to educate politicians and influence public opinion.

This book is divided into five parts:

Part I: Getting Online familiarizes you with the basics of telecommunications. It describes what computer hardware and software you need. It explains how your computer transmits data over the phone lines and how to set up your system so you can get online immediately and without fuss. If you are already familiar with networking, you can skip this section.

Part II: Global Networks explores four massive global networks that link millions of individuals around the globe: FidoNet, BITNET, Usenet, and the Internet.

Part III: Electronic Bulletin Boards introduces you to a fascinating world of thousands of independent, and often idiosyncratic, bulletin board systems (BBSs).

Part IV: Commercial Online Services tells you about the environmental resources that can be tapped through the online networks—such as America Online,® CompuServe, EcoNet, GEnie, and The WELL—that have been developed and marketed to meet the needs of the increasing numbers of individuals who have personal computers in their homes.

Part V: Libraries That Never Close shows you how you can save hours of research time—and numerous trips to the library—by searching through vast bibliographic databases from your home computer. This section also examines the use of CD-ROM technology for environmental research and describes the online services that bring you "hot-off-the-press" environmental news. Using these databases, you can gain access to almost any environmental or scientific article or publication in existence.

Throughout the book, the focus is on how to get online and on what types of information and people you can find online. This volume concentrates on resources pertaining to natural and physical sciences. Those services that cover politics, the social sciences, humanities, medicine (other than public health), and other disciplines are outside the scope of this book, although a few special cases have been included. If you are interested in those topics as they relate to environmentalism, however, you can find them in the networks and services outlined in this book.

Environmentalism is a multi-disciplinary pursuit, and this book introduces only a small subset of the vast resources available online. As you explore those resources, you will likely find yourself being led to other areas of interest, even those unrelated to the environment. Once you become comfortable online, exploring is part of the reward.

The various online services described in this book are not all equally as easy to use. Many of the global networks, for example, were created by technical programmers in academia, government and business; these networks' user interfaces differ greatly and often require that you memorize complex technical commands. The commercial online services, on the other hand, were developed specifically for a mass audience and tend to be easier to use. You will find, however, that ease of use also varies among the commercial networks; some mimic the academic mainframes, while others use simple "point-and-click" navigation schemes.

Getting Help

I have attempted to show you how to get online as easily as possible, but if you need help, consider the following:

- Almost every city in the country has a local computer user group. User groups are friendly groups of people who are in the business of helping other computer users. You will have little difficulty finding a member who will gladly help you learn the ABCs of telecommunications. Just about any computer or software retailer can put you in touch with a user group in your area. *Vulcan's Computer Monthly* (P.O. Box 55886, Birmingham, AL 35255; 800/874-2937) each month publishes a state-by-state list of user groups along with the computers they support. If you use a Macintosh or other Apple computer, call Apple Computer toll-free at 800/538-9696, extension 500, and ask for the name and contact of an Apple user group in your city.

- Most of the commercial online services and databases described in this book have a toll-free number that you can call for assistance in getting online for the first time.

Getting online is only the first step. Once you have started your online adventure, you will have no difficulty finding additional avenues of exploration:

- *Vulcan's Computer Monthly* features a monthly list of bulletin boards across the country that are available to the public. The list may help you locate a bulletin board in your city.

- Hayes Microcomputer, a leading manufacturer of modems, features an online technical bulletin board (800/874-2937) from which you can obtain information on Hayes' products. Hayes' bulletin board also provides the complete Darwin's bulletin board list, which lists hundreds of bulletin boards in the United States. Since you can use keywords to search the database, select your area code to locate bulletin boards in your area quickly.

- The MNS Online BBS (518/381-4430) is available 24 hours a day, seven days a week, and offers many public message boards, private email, and file sections for downloading. It is free, is user friendly, and has a special section for EcoLinking established by the author of this book.

- Many of the bulletin boards featured in this book provide lists of their own favorite bulletin boards.

Once you get comfortable being online, you can study the more technical aspects of telecommunications by reading books and articles listed in Appendix E. I highly recommend three books that cover global networks in great detail: *The Matrix: Computer Networks and Conferencing Systems Worldwide,* by John Quarterman; *A Directory of Electronic Mail, Addressing and Networks,* by Donnalyn Frey and Rick Adams; and *The User's Directory of Computer Networks,* by Tracy L. LaQuey (see Appendix E).

If you know of a bulletin board, a network, or a service useful to the environmental community, let me know by sending me email through one of the following services: America Online (AFL DonR); GEnie (MNS); CompuServe (70057,1325); BITNET (drittner@albnyvms); Internet (drittner@uacsc1.albany.edu); AppleLink (UG 0194); EcoNet (drittner); or MNS OnLine (518/381-4430).

Knowledge is the only instrument of production that is not subject to diminishing returns.

J. M. Clark,
Journal of Political Economy, 1927

The Basics

This chapter teaches you the basics of getting online. If you are already familiar with how to do so, skip this chapter. Go ahead and explore the various networks, bulletin boards, and databases featured later in this book.

This chapter introduces the basic equipment you need to get online:

- a computer with a serial communications port,
- a "Hayes-compatible" modem,
- a communications software package, and
- a phone line.

We will look at the basic elements of "data," computing jargon for the information that you transmit from your computer to the computers of others and vice versa. Finally, you will be introduced to the basic terminology of telecommunications.

Your Computer

Let's assume that you are already familiar with your personal computer. It doesn't matter if you have an Apple II, a Macintosh, an Amiga, a Commodore, an Atari, an IBM, a Tandy, a Compaq or another type of IBM compatible. Most of those (except Apple II and some IBMs and IBM compatibles) have built-in communications ports for telecommunications. If your computer does not have a built-in serial port, see your computer dealer for an inexpensive upgrade.

Consider taking an introductory computer course at your local college, university, or continuing education center if you have never used a computer before. Or you may want to join a local user group. User groups frequently offer computer courses for beginners at little or no charge.

If you know how to type (even if only with two fingers), all you need to do is turn on a switch and follow some basic directions and you're ready to EcoLink. Simply put, you place a communications software package on your computer. Then the computer, hooked up to a device called a modem, uses the phone line to call a bulletin board, a network, or an online service to send and receive the information you wish to share with others (the "data").

Communicating with a Modem

Data and Serial Ports

Asynchronous telecommunication is the name given to the type of communication described in this book. It's a fancy way of saying that the information you send over the phone line may not transmit at regularly scheduled intervals.

During asynchronous communications, the serial port in the back of the computer is very busy. The serial port is responsible for making sure your data is translated correctly by "talking" to your modem.

Your data, be it a report you type, a picture you draw, or a number in a spreadsheet, is composed of tiny electrical pulses called *bits* (short for *binary digits*). Each represents a single digit of data and is either a one or a zero. Think of it as an electrical switch, with one being on, and zero being off.

The bits are transmitted as electrical signals by your computer to a modem that is plugged into the serial port. A modem is a device that transforms digital data into sound and sends it over the phone line to a receiving computer (that is, a bulletin board, another computer, or an online service).

The serial communications port at the back of your computer has either a 25-pin or a 9-pin connector. The ports are often labeled *serial, RS-232, comm,* or *modem.* In the case of some Macintoshes (those in the Plus to Mac II series), you will find a miniature 8-pin port, indicated by a picture of a phone. In some models, like the Apple II series and some IBM compatibles, you need to buy a serial interface card and install it in the computer.

The pins on the serial port use a change in voltage level to transmit each data bit. The "one" bit transmits at one voltage level, and the "zero" bit transmits at a different voltage level.

When you type text onto the screen, the computer breaks it down into its binary form—the combinations of ones and zeros—that represents the characters you see displayed. To ensure consistency, each character is based on a standard called the ASCII (American Standard Code for Information Interchange) character set. The ASCII code ensures that each letter of the alphabet has its own ID, in a pattern of eight bits. So, for example, the lowercase *a* in its binary form is *1100001*, and the uppercase *A* is *1000001.* (ASCII uses only seven of the eight bits; the last one is always a zero.)

This group of eight bits is called a *byte*. Bytes are measured in units of a thousand—*kilobyte.* Since each byte represents a single character (either a letter, a number, or a symbol), a typical computer system with one megabyte of memory can store more than a million bytes, or about 180,000 words. A single 3½-inch floppy disk can hold more than 200,000 words.

Modems

A modem is a device that connects a computer to a telephone line and converts the digital data from the computer to analog (sound) frequencies. The modem sends the sounds through the phone line to a receiving computer's modem (see Figure 2-1).

The receiving modem in turn changes the sounds back into a digital form, and the data is displayed on the receiving computer's screen. This two-way process is called *modulation/demodulation*, and the word *modem* is actually a contraction of modulator/demodulator.

Modems that send data in both directions simultaneously are called *duplex* modems. This type of modem is used for the communications described in this book.

FIGURE 2-1

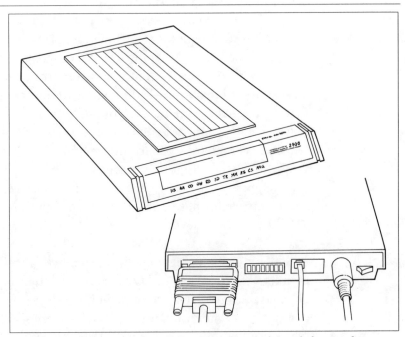

Drawing at bottom shows the back of the modem: (left to right) port that connects the modem to the computer, dip switches, phone line connection, power cord, on/off switch.

The speed at which modems transmit data over phone lines is measured in bits per second (bps) and is also known as the *baud rate*. The higher the bits per second number, (typical rates are 1200 bps, 2400 bps, 4800 bps, 9600 bps), the faster the transmission of data.

It wasn't long ago when 300 bps was the only speed possible for telecommunications. Today, however, 2400 bps is the most popular speed, with 9600 bps fast becoming affordable. Do not buy a modem that transmits below 1200 bps, as most bulletin boards now refuse to grant access to anyone using 300 bps. Most higher speed modems do transmit at the lower speeds if necessary. In general, the higher the baud rate of a modem, the more it will cost; however, with the cost of modems coming down all the time, you will likely find a 2400 bps or faster modem that falls within your budget.

All modem manufacturers have adopted standards, based on early AT&T standards, for 300 bps and faster modems. This standardization allows data transmission among all those modems. Be careful about buying modems that transmit at 19.2k bps, since no standards are currently agreed upon for them. Moreover, since standards for 9600 bps modems have been adopted only recently, 9600 bps modems that do not work properly may be available in the market place. To be on the safe side, look for a modem that says it meets the CCITT V.32 standard. CCITT stands for Comite Consultatif International de Telephonie et Telegraphie; this international committee sets the standards for the interconnections between modems and computer terminals. The standards start with the letter *V*, for example, V.21 is the standard for 300 bps, V.22 for 1200 bps, V.22bis for 2400 bps, and so on.

Modems come in two models, external and internal. Most modems today are called "smart modems" because they have a built-in language command set that performs functions dictated by the computer. Hayes Microcomputer set the standards, so make sure the modem you purchase says *Hayes-compatible*, meaning compatible with the *AT* command set. AT commands are sent out from the computer's serial port and tell the modem what actions to perform, such as answer the phone or dial a number.

For example, when a Hayes-compatible modem is dialing a number, the display on your screen reads `ATDT[the phone number]`. The *AT* command gets the modem's "attention." This code precedes all command lines. *DT* means "dial touch-tone;" if you have a rotary phone, the command would be *DP* for "dial pulse."

While modem manufacturers boast that their modems offer many bells and whistles, the most important functions are the abilities to automatically dial and answer a computer. Additionally, besides being compatible with the Hayes command set, the modem should automatically sense the speed of another modem and drop down to match—important when you call a BBS that has a lower speed than yours—and should send and receive faster than 300 bps. You will find many models from which to choose; your favorite hardware retailer or computer magazine will describe the advantages and disadvantages of each one.

Cables and Connections

You connect the serial port of your computer to the modem with a special cable. It is important to make sure you get a cable that is correct for your brand of modem and computer. If you spend the few extra dollars it takes to buy a cable from the modem manufacturer you can be sure you have the correct cable. Some modem manufacturers—for example, Hayes—now make the cable part of the modem (as with the Hayes Personal Modem series) or include the cable with purchase of the modem.

Communications Software

You need special software to enable your computer to communicate with your modem. Once you select and set up your communications software package, most of the communications functions are performed automatically by the computer and the modem (see Figure 2-2).

Many software choices are available for every model of computer. Some are commercial products costing a couple hundred dollars and some are available free from user groups and bulletin board services. Some of the free programs have as many features as, if not more features than, the commercial products. Appendix A lists a number of popular communications software programs.

What should you look for in a software program? Make sure you can easily change parameters such as the modem port, the baud rate, the parity, and the stop bits. Additionally, the soft-

FIGURE 2-2

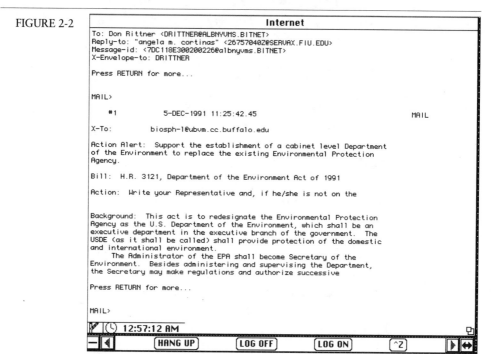

Communications software programs are "windows" that allow you to log on to bulletin boards. Here MicroPhone (a Macintosh software program) is logged on to a local university mainframe through which I have access to Internet and BITNET.

ware should have two or more error-checking protocols such as Xmodem and Ymodem (the protocols all have ASCII download capability). Make sure you can select terminal emulation, the ability to mimic other computer types, so you can log on to mainframes (common on some of the big academic networks). The ability to store phone numbers allows you to call your favorite bulletin boards without typing in the phone number each time. Several packages allow you to write *macros* or *scripts*—shortcuts that automate signing on, signing off, or other procedures like downloading and uploading files. Remember to purchase communications software written for your type of computer.

Communications Jargon

You need to be aware of a few extra communications functions so you can set up your software properly to communicate with a bulletin board service or a network.

Start Bits and Stop Bits

Because of the varying time factor in asynchronous data transmission, it is necessary to add an extra bit at the beginning and the end of each byte as it is transmitted. The extra bits are called *start bits* and *stop bits*.

The start bit tells the receiving computer that the next bit begins the data flow. The stop bit tells the computer it received the complete byte. Most bulletin boards and networks require your communications software to be set to one stop bit; others will ask for two stop bits. Either way, you need to know how many. Most bulletin board listings require one stop bit, so when in doubt, set up your communications software that way.

Parity

Parity refers to the number of ones in a binary number. If the quantity is even, the number has "even parity"; if the quantity is odd, then the number has "odd parity." Parity checks for errors in transmission. If parity is used, a one or a zero is placed right before the stop bit, depending on what type of parity you select—even or odd—and on the parity of the binary number in the data byte being transmitted.

So if you select even parity, and the byte contents have odd parity, the parity bit will be a one, and the parity of the entire byte will be even. Conversely, if you select even parity and the data byte already has even parity, the parity bit will be zero, and even parity will be maintained. Why do you need to know this? The sending and receiving computers must have the same parity.

Ninety-nine percent of the bulletin boards and services described in this book require you to set your software parameters to "8-0-1": eight data bits, no parity, and one stop bit. If you set your software to those parameters, you should have no trouble getting on the services outlined in this book.

Duplex

Duplex and *half duplex* are terms that refer to the direction of data transmission. Half duplex means data is sent between two computers one direction at a time. Duplex, or full duplex, means the data is transmitted in both directions simultaneously. Some error-checking protocols like Xmodem are half duplex. The sending computer must wait for the receiving computer to confirm that it received each block of data before the first computer sends another block. This often slows down transmission, since line noise on the phone can disrupt transmission, and the block of data must be retransmitted. Most bulletin boards and services covered in this book require a duplex setting.

File Transfer

This book describes bulletin boards, networks, and services that contain a wealth of environmental and scientific information. An important part of being able to use that information is being able to transfer it to your own computer—a process called *file transfer*. The information you get from online resources can be in the form of text (ASCII) files or binary files (programs, applications, graphics, spreadsheets, or formatted ASCII files). *Downloading* is the process of taking information from a bulletin board or a network. *Uploading* is the process of placing information stored on your personal computer onto a bulletin board or a network for others to use.

Your communications software allows you to download information using several protocols to make sure the resulting download is error free. You can capture text documents or straight ASCII text right from the screen. Most software programs let you capture the text as it scrolls down the screen and save it into a text file. You can also download the file using the ASCII download command; however, ASCII contains no error checking, so if you have a bad phone line, the resulting text download may be jumbled beyond recognition.

Most communications programs carry two or more error-checking protocols. The most commonly used downloading protocol is *Xmodem*. After sending each block of 128 bytes, Xmodem checks the accuracy of the transmission, telling the other computer to resend the block if it contains errors.

Xmodem-crc is a version of Xmodem that uses a more accurate method of error checking called Cyclic Redundancy Check.

Ymodem uses larger blocks than Xmodem—1024 bytes—and allows *batch transfer,* or the ability to download more than one file at a time.

Kermit is the name of both a protocol and a communications program. Kermit is popular for downloading from mainframes and minicomputers on college campuses. It's slow, but you can use it for batch transfer.

Zmodem, the newest protocol, also downloads multiple files. In addition, it can recover its transfers intact if you temporarily lose the connection (see Figure 2-3).

The only thing you need to remember is that both computers must use the same protocol when you are downloading (to your computer) or uploading (to the other computer). That's why

FIGURE 2-3

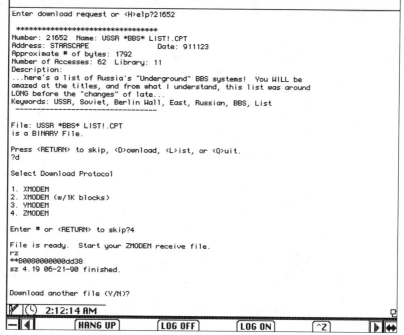

Downloading a file from GEnie using the Zmodem protocol.

when you select your communications software package, you should select one that supports a number of file-transfer protocols. Nine out of ten times, however, you will use ASCII or Xmodem file transfers on the services discussed in this book.

Archiving

Several programmers have created *archiving* programs that let you compress a file or a group of files into one smaller file. This makes files easier (read "faster and less expensive") to download or upload (see Figure 2-4).

If you download an archived file, you need to have the archiving program on your own computer to open the file. Almost all archiving programs have a free or shareware version to

FIGURE 2-4

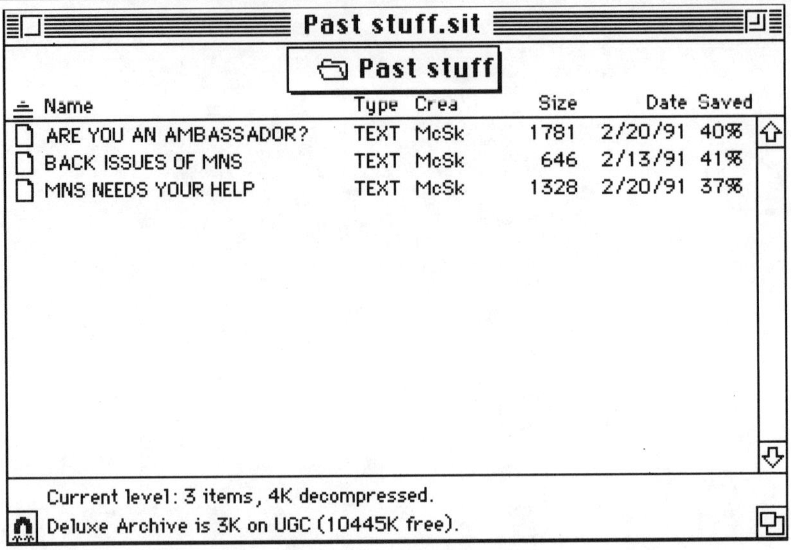

```
┌─────────────────────────── Past stuff.sit ───────────────────────────┐
│                          📁 Past stuff                                │
│  Name                        Type  Crea    Size    Date   Saved       │
│  📄 ARE YOU AN AMBASSADOR?   TEXT  McSk    1781   2/20/91  40%        │
│  📄 BACK ISSUES OF MNS       TEXT  McSk     646   2/13/91  41%        │
│  📄 MNS NEEDS YOUR HELP      TEXT  McSk    1328   2/20/91  37%        │
│                                                                       │
│  Current level: 3 items, 4K decompressed.                             │
│  Deluxe Archive is 3K on UGC (10445K free).                           │
└───────────────────────────────────────────────────────────────────────┘
```

Archiving programs like StuffIt, shown here, allow you to compress several files into one, decreasing the time it takes to upload or download them from a bulletin board.

unarchive files. (*Shareware* means you get the program free and pay the programmer if you use it.)

You can identify an archived file because an extension abbreviation appears after its file name. For example, you find ZIP, ARC, PAC, ZOO, and LZH in the PC world and SIT, DD, STF, PKIII, and CPT in the Macintosh world. All the bulletin boards mentioned in this book provide you with the unarchiving utilities for the file types that are on their boards.

Virus Checking

Before you download any file into your computer, you should have a program to check for "viruses" that can harm your data or your hardware. Many "antivirus" programs are available, both free and commercial, that do the trick. Some automatically check any file that you place on your hard drive or insert into

FIGURE 2-5

```
for viruses, remove any viruses which you may have on
your system, and protect your system against future
infections. To read the manual, select the "Disinfectant
Help" command in the Apple menu.

================================================

DONS40
Disk scanning run started.
12/6/91, 2:15:40 AM.

### Scan canceled.

================================================

E#2
Disk scanning run started.
12/6/91, 2:16:09 AM.
Disk scanning run completed.
12/6/91, 2:16:13 AM.
13 total files.

No infected files were found on this disk.
```

Files scanned: 36
Infected files: 0
Errors: 0

Disinfectant, an excellent and free virus checker by John Norstad for Macintosh computers.

your floppy drive. Some are applications that you need to run periodically.

The best place to obtain virus checkers is your local user group or online service. All the commercial online services, such as America Online, CompuServe, and GEnie, check for viruses before they place files online for public distribution (see Figure 2-5).

Saving on Phone Costs

Calling bulletin boards around the country can add up on your phone bill, but there are a number of ways to minimize the cost. Most of the large online services such as America Online, CompuServe, and GEnie provide local telephone access using public data networks (known as *packet switching networks*), such as SprintNet (formerly known as Telenet), Tymnet, and in Canada, DataPac. These services transmit the information that you and other computer users send back and forth by breaking it up into "packets" of data and then transmitting the packets across dedicated high-speed transmission lines.

US Sprint and the Galaxy Information Network also provide low-cost ways to gain access to the many bulletin boards found in large metropolitan areas of the United States. US Sprint offers a service called PC Pursuit that reaches 34 of the country's major cities; access to those cities is a local call from more than 18,000 cities. PC Pursuit has a one-time registration fee of $30 and costs $30 a month for 30 hours of non-prime-time use (weekends and 6 p.m. to 7 a.m. weekdays). Family memberships (60 hours for $50 per month) and handicapped memberships (90 hours for $30 per month) are also available. If you pass the 30-hour allotment, you pay an additional charge of $3 per hour. You pay $10 per hour for prime-time use. Call 800/736-1130 to get more information about PC Pursuit. You can find the local SprintNet node in your region by calling Telenet at 800/TELENET.

Galaxy StarLink uses the large BP Tymnet data network, owned by British Telecom. You call a local Tymnet node and enter your password and the city's destination code. Once you are connected to that city, you dial the BBS number as if it were a local call. StarLink has a one-time $35 registration fee, a $10 per month maintenance fee, and a non-prime-time rate of $1.50 per hour to transfer up to 200K (kilobyte) per hour. Non-prime time is weekends and 7 p.m. to 6 a.m. weekdays. To find out more about Galaxy StarLink, call 505/881-6988.

What Have You Learned?

- All this complex computer stuff is really nothing more than bits of electrical pulses going on and off and being displayed on a screen in plain English (or whatever your language may be).

- You need to have a serial port in your computer, a modem, a communications software package, and a phone line.

- Your modem must be Hayes compatible, have automatic dial and receive functions, and at least 1200 bps.

- The cable that connects your computer to your modem must have the right pin configuration, and it needs to be wired correctly. Your safest bet is to buy a cable manufactured specifically for your brands of modem and computer.

- You need to set the communications parameters correctly to match the computer you are calling. Additionally, modem speed must be the same on both computers, and you must match the baud rate, parity, stop bit, and duplex.

- Many communications software programs are available for your computer. Some are free or low-cost, and some are easier to use than others. Ask for recommendations.

- You can use one of many error-checking protocols to download files into your computer. Be sure that the communications program you buy supports at least ASCII and Xmodem file transfers.

- Several programs are available to archive or pack one or more files, and to decompress them, making it less expensive to upload and download.

- Never download a file into your computer without having a program that checks for viruses.

Since every communications software program is different, it is impossible to describe here exactly how to set up your communications software and dial another computer. Follow the instructions that come with your software.

Online Etiquette

As you communicate with other people online, you should follow a few rules of etiquette. Heeding the following pointers will make your online adventures productive and fun.

- DO NOT TYPE YOUR E-MAIL OR MESSAGES IN ALL CAPS! THIS MEANS YOU ARE SHOUTING! (Exceptions to this rule are if you really do mean to shout and if your computer only takes input in capital letters.)

- Remember that another person is on the receiving end of your e-mail or public message. The same social graces that apply in everyday life also apply to online communications. Do not attack other people if you cannot convince them to accept your opinions or ideas. Screaming (using all caps), using profanity, and abusing others are not acceptable behavior.

- Be succinct in your postings, both private and public. If you can't make your point in 50 words or less, re-think how to word it.

- Think first, write later. Before you respond emotionally to someone's posting, think your answer through. Once your response is posted, it becomes part of the "virtual world," and you cannot get it back.

- Be careful about using subtleties and sarcasm. They do not translate well into online communications and can lead to misunderstandings.

- Respond only to a public posting if you have something substantial to say. Wading through hundreds of answers such as "Yeah, right on," and "I agree," do nothing except waste the reader's time.

- If you are posting material from sources, be sure to give credit where credit is due. Cite copyrights and licenses. Violating someone's copyright or license could cause you legal hassles.

- Use descriptive titles when you post a message. (This helps "browsers" make the right selections.) If your title is "Macintosh Forever," you could be a computer user or a zealous apple grower.

- If you download lots of public-domain material from a BBS, give something back, be it new files or messages.

- Stay on topic. If you are on a bulletin board that specializes in botany, asking questions about environmental impacts in space will likely get you little help.

- Summarize. When you respond to someone's public posting, summarize the parts you are responding to. This allows readers to understand your comments instead of trying to remember what the original post was about. Some software allows you to insert the text you are responding to automatically by placing a caret in front of the comments.

- Don't post business offers, chain letters, or other entrepreneurial adventures on a public message board unless the board was created for it. By the same token, do not send unsolicited commercial messages (junk mail) via e-mail. Nothing is more annoying to readers than a long tirade about how you can make millions for no money down.

- Avoid conflicts, by identifying yourself. For example, if you are discussing the value of a piece of software or hardware and you work for the company that produces it, inform your readers of this fact.

Smileys

Since it is impossible to notice vocal inflections or emphasis when you communicate online, a special shortcut for depicting emotions has evolved: "smileys." Smileys help you communicate what words alone cannot express in this "faceless" world, and they add an element of fun and levity.

By turning your head to the left, you can recognize the characters. For example, this is a smile ➔ :-) (two eyes, a nose, and a mouth; the nose is optional.) The following is a sample of the hundreds of smileys you will encounter while networking. Good luck! :)

:-o	Wow!	:-(	Frowning	;(	Crying
:-|	Grim	:-*	Kiss	[]	Hugs (with
:-v	Speaking	'-)	Wink		recipient's
:-,,	Smirk	;-)	Wink		initials in
:-||	Anger	:-[	Pouting		brackets)
:-)	Smiling	:-#	My lips are	}:-(	Bullheaded
:-p	Sticking out		sealed	:-&	Tongue-tied
	tongue	:O	Yelling		

Some abbreviations used in electronic mail and in "live" conferences include the following:

LOL	Laughing out loud	/	Done (after making a remark)
GMTA	Great minds think alike		
IMHO	In my humble opinion	ROFL	Rolling on the floor laughing
BTW	By the way		
BRB	Be right back	...	More to come
GA	Go ahead	?	Want to ask a question
		!	Want to make a comment

PART II

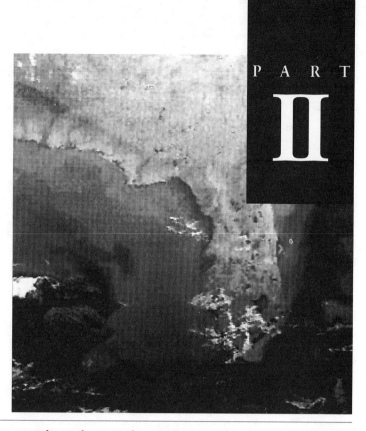

Global Networks
Worldwide Environmental Connections

Four computer networks belong to what writer John Quarterman calls the "Matrix," a worldwide meganetwork of connected computer networks and conferencing systems. The four networks—FidoNet, BITNET, Usenet, and the Internet—electronically connect millions of scientists, environmentalists, business people, students, professionals, and researchers in more than 40 countries.

These global networks contain vast environmental, academic, bibliographic, and scientific resources, combined with the brainpower of millions of individuals who have the power to communicate online. This powerful combination forms what could be called the *GaiaNet*—a powerful econetwork serving the worldwide environmental community. You can become part of the GaiaNet by participating in any one of those four networks, since they allow the transfer of electronic mail ("e-mail") and messages among them.

A word of warning: The global networks were designed by scientists, programmers, and other technically minded people, and the commands needed to master them were not designed for ease of use. Also, since the networks are not all based on the same operating system, you may find a frustrating lack of consistency as you explore.

As the power of telecommunications is discovered by increasing numbers of nontechnical people, leaders within the global networking community are taking steps to make the networks more "user-friendly" in design. In the meantime, the vast resources available through these networks—combined with the tremendous power associated with being able to communicate with literally millions of individuals around the world—make it well worth your while to explore this alien territory.

FidoNet

FidoNet consists of several thousand individual bulletin boards that share mail and conferences. FidoNet is open to anyone with a computer and usually carries no fee. You can find participating bulletin boards in almost every city in the United States, so it's a local phone call away for most people.

BITNET

BITNET (Because It's Time Network) connects academic institutions in more than 30 countries. You need affiliation with a "host" site (your local university, for example) to participate. On BITNET, you can subscribe to "mailing lists"—discussion groups on various subjects—send mail, and even have short interactive discussions with fellow users.

Usenet

Usenet (User's Network) is a series of conferences, called *newsgroups*, participated in by a worldwide network of computer users. Usenet is distributed on computer networks running the UNIX operating system, and you can find the network at universities, businesses, and government and military sites. In addition, a few commercial online information services and bulletin board services carry Usenet newsgroups. You can obtain a Usenet account, usually for free, at a local university. On commercial online services, you often pay a fee to participate. A few free public UNIX sites and several FidoNet bulletin boards act as gateways to the Usenet newsgroups.

Internet

The Internet is a meganetwork of more than two thousand networks. Subsidized by the U.S. Government, the Internet provides access to supercomputers, e-mail, bulletin boards, databases, many of the country's best library catalogs, and discussions with many of the leading scientific minds. Much of the Internet is not open to the public, but access to the National Science Foundation's NSFNET and affiliates is possible for research purposes.

> The notion that only the short-term goals and immediate happiness of Homo sapiens should be considered in making moral decisions about the use of Earth is lethal, not only to nonhuman organisms but to humanity.
>
> Paul Ehrlich
> *Extinction*, 1981

CHAPTER 3

FidoNet

FidoNet—known as "the people's network"—is an amateur network of more than 10,000 individual bulletin board services (BBSs) electronically connected across six continents.

In 1983, computer programmer Tom Jennings began to develop bulletin board software that he called *Fido*. Jennings had no idea that he was about to create one of the most important products of the communications revolution. He was intrigued by ongoing discussions in the computer world about creating a communications link between the East Coast and the West Coast using homegrown bulletin boards—a scheme loosely patterned after the amateur ham radio operators' network. Jennings' original scheme was for a bulletin board system on the East Coast to call another local bulletin board automatically. The second bulletin board would call a third local bulletin board, and so on. Eventually, using only local calls, the message would reach the receiving bulletin board on the West Coast. While it was a good idea on paper, it wasn't practical in operation, so Jennings developed special software that would do the job. His vision of affordable, coast-to-coast communication resulted in FidoNet.

Jennings was obviously on the right track. FidoNet grew from two bulletin boards to over 10,000 in little more than five years.

Typical bulletin board operations include private and public mail and file libraries. FidoNet expands on those services by offering three additional ones: Netmail enables you to send private messages to any participant on the FidoNet; Echomail consists of public conferences on a variety of topics; and file transfers allow you to send reports, software programs, spreadsheets, or other files, to anyone on the FidoNet.

Echoes are the "meat and potatoes" of the Fido world. Here you will find discussions on a wide range of topics, from how to bake an apple pie to the origins of the universe, ecological disasters, or the latest developments in AIDS research. Such a diversity of subject matter and opinion exists on the Fido network that it is truly the "people's network"—the heartbeat of world discussion of the issues facing us today.

FidoNet is a good place to get your feet wet if you're going online for the first time. Fido *SysOps* (system operators) are friendly and cooperative and will guide you through their bulletin boards (sometimes just called *boards*). With thousands of locations across the United States, chances are a Fido BBS operates in your own city, saving you long-distance charges while you learn.

How to Get Online

The original Fido network design was for MS-DOS computers. However, other systems, such as the Apple Macintosh, the Amiga, and the Atari ST, soon developed Fido capability and are now part of the network.

To participate in FidoNet, you must dial into a BBS that is a part of the FidoNet. For the number of the host nearest you, call the MNS OnLine BBS at 518/381-4430 and download the latest copy of the Fido "Nodelist." (You may also find this list on a number of other bulletin board services.)

Cost

If you find a BBS that belongs to FidoNet in your area, access will cost you only the price of a local call in the majority of cases. The entire cost of maintaining the network is distributed among the participating SysOps. Many boards allow you to send private mail for a small fee, or charge you to enter the Echo conference area. However, most boards are free, so be kind to your SysOp.

FidoNet Organization

The Fido world, while anarchistic, does have enough order to it to make it work, and many dedicated people help keep it together. Since the past year has seen a dramatic increase in the number of new participants from around the world, a movement is underway to approve a worldwide FidoNet policy. The Fido world breaks down into Zones, which break down further into regions, networks, hubs and nodes. See Figure 3-1. **Zones** are broad geographic areas that may include more than one country or continent. The United States and Canada are both located in zone 1, for example (see Table 3-1).

TABLE 3-1 Fido Zones	Zone	Location
	1	North America
	2	Europe
	3	Oceania
	4	Latin America
	5	Africa

Smaller geographic areas are called **regions** (see Table 3-2). For example, region 13 in zone 1 comprises all the Fido boards in Washington, D.C., and the states of Delaware, Maryland, New Jersey, New York, Pennsylvania, Virginia, and West Virginia. Each region is presided over by a regional coordinator, who maintains a list of member nodes. Each regional coordinator periodically sends the regional node list to the international

FIGURE 3-1

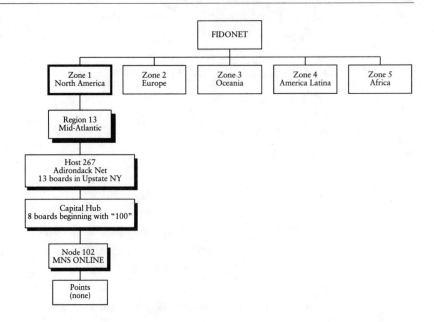

coordinator, who compiles the master node list—the Fido bulletin board directory. The master node list is updated each week and sent to every node on the network, so nodes can send and receive mail among themselves. The weekly update file is called the Nodediff.

TABLE 3-2 FidoNet Regions	Region	Location
	Zone 1:	**North America**
	10	West: Calif., Nev.
	11	Central: Ill., Ind., Ky., Mich., Ohio, Wis.
	12	Eastern Canada: Ont., Que., N.B., N.S.
	13	Mid-Atlantic: Del., D.C., Md., N.J., N.Y., Pa., Va., W. Va.
	14	Midwest: Iowa, Kans., Minn., Mo., Nebr., N. Dak., S. Dak.
	15	Mountain: Ariz., Colo., N. Mex., Utah, Wyo.
	16	New England: Conn., Maine, Mass., N.H., R.I., Vt.

17   Northwest: Alaska, Hawaii, Idaho, Mont., Oreg., Wash., B.C., Ala., Sask., Man.
18   Southeast/Caribbean: Ala., Fla., Ga., Miss., N.C., S.C., Tenn., P.R.
19   South: Ark., La., Okla., Tex.

Zone 2: Europe
20   Sweden
21   Norway
22   Finland
23   Denmark
24   Germany
25   British Isles
28   Holland
29   Belgium
30   Switzerland
31   Austria
32   France
33   Italy
34   Spain
35   Bulgaria
40   Israel
41   Greece
42   Czechoslovakia
48   Poland

Zone 3: Oceania
50   Australia
51   Asean (Singapore)
53   Hong Kong & Macao
54   Western Pacific
55   The Million-Square-Mile Region
56   Far East
57   New Zealand

Zone 4: Latin America
80   Brazil
90   Argentina

Zone 5: Africa
48   Botswana
49   South Africa
72   Zimbabwe
73   Kenya

Regions break down into **networks**, called *nets*. Each net has a host, usually the network coordinator's bulletin board. The network coordinator is the SysOp who collects mail, routes it, and maintains a list of all the nodes in a network. More than 200 nets exist in the United States and Canada alone. You can usually call the host near you by modem and request the number for the local Fido node in your city. Longevity of BBSs vary, so the number of hosts and Fido members constantly changes. Nets break down into **hubs**—even smaller geographical areas—which may contain one or several individual bulletin boards. Lastly, each individual BBS that belongs to FidoNet is a **node**, the smallest unit in the FidoNet.

Each BBS that belongs to the Fido network has a unique address. Let's look at an example. I maintain the MNS Online BBS in Albany, New York. As part of the Fido network, the board's address is 1:267/102. This number signifies that the board is in Zone 1, North America, and net 267, the Adirondack net. This net consists of 13 bulletin boards, eight of which make up the Capital hub. Node numbers in this hub are composed of three digits and begin with a one, the individual board in this example is node 102. So anyone who uses FidoNet can send me a message at my Fido address—1:267/102—from anywhere in the world at the cost of a local telephone call. Figure 3-1 shows where the MNS Online BBS falls in the FidoNet hierarchy.

Gateways to Other Networks

Tim Pozar and his colleagues have written software called UFGate that allows the FidoNet to share mail and discussions with the Internet, BITNET, and Usenet.

This gateway to other networks is an exciting development, since it broadens the reach of FidoNet users, connecting them with much of the leading discussion and research in the scientific and environmental fields. Several Fido boards now gateway certain echoes to BITNET and the Internet.

FidoNet mail to the Internet is routed through the various FidoNet BBS gateways. FidoNet administrator David Dodell explains how to do this:

> **Internet via FidoNet**
>
> FidoNet is fully coupled into the Internet. You do not need to know any specific gateways to send mail from FidoNet to Internet; all you have to do is address the message correctly into the Fidonet.org domain, and the message will be routed automatically. Here's how to address messages through FidoNet to the Internet: The basic FidoNet address format is:
>
> ```
> FirstName.LastName@pww.fzz.nxx.zyy.fidonet.org
> ```
>
> ww = Point number (this is usually not needed unless specific to a subsystem)
>
> zz = FidoNet node
>
> xx = FidoNet network or region
>
> yy = FidoNet zone
>
> As an example, my name David Dodell, resides at FidoNet address 1:114/15. My FidoNet Internet address is:
>
> ```
> David.Dodell@f15.n114.z1.fidonet.org
> ```

Now, how do you go from a FidoNet node to an Internet-style address? It's just as easy; however, you need to find a gateway on FidoNet first, since no automatic routing to Internet gateways is available at this time. For example, you could use my gateway at 1:114/15. You would send a message to the user "uucp" at 1:114/15.

In the first line of text, you put the Internet-style address, followed by two returns:

```
To: user@site.domain
```

For example, to send to my BITNET account of ATW1H@ASUACAD, the FidoNet message would go to "uucp" at FidoNet address 1:114/15. The first line of text in the body of the message would be

```
To: atw1h@asuacad.bitnet
```

Or, to my uucp address:

```
To: postmaster@stjhmc.uucp
```

Environmental Echoes

Special public topics or conferences on the FidoNet are called *echoes*, short for *Echomail*. About 400 echoes now function on the Fido network. They are generally distributed over the "backbone," the nodes that shoulder the bulk of the traffic. New echoes are constantly created, and unpopular ones come and go. Each echo has a moderator, usually the person who originally proposed the echo.

Members of an echo do not communicate with each other in "real time"—that is, they are not necessarily on FidoNet at the same time. Instead, each comment and response from a member is distributed to every other member of the echo to be read and in turn responded to at each member's leisure.

Dennis McClain and the SEEDS Project

Newly single, Dennis McClain had just moved to a new town and had more time than he knew what to do with. At the urging of friends he met at a local Apple computer user group meeting, Dennis bought a modem.

Within two months, Dennis was a regular on his user group's BBS and on a commercial system based in his town. Within two more months, he was the system operator of the club's BBS and an assistant operator on the commercial system.

Soon Dennis was spending so much time online that in order to keep from missing telephone calls, he put in a separate phone line for his modem. "I think this is the first sign of a serious telecommunicator—a dedicated modem line," he says.

When Dennis first approached FidoNet, he had little understanding of how to navigate and use it. But within hours, he realized that it represented an enormous force in communications.

Dennis's interest in the space program led him to FidoNet's Space Development echo, moderated by the SysOp of the National Space Society's BBS. Here, Dennis heard about the SEEDS Project, in which 12.5 million tomato seeds that had orbited the earth for six and a half years were to be distributed to schools across the country. One of his own sons was in a class that was to receive a package of the seeds. Although Dennis had planned to provide the school with information he received on the project, it occurred to him that perhaps the students would like to connect to FidoNet themselves.

Dennis approached NASA with the concept of connecting students involved in the SEEDS Project worldwide through FidoNet. After several meetings and demonstrations, NASA agreed to cooperate. With the help of the Department of

> Education's BBS, schoolchildren all over the world were soon linked through FidoNet. Although the SEEDS Project is over, the telecommunications links are still in place in schools worldwide.
>
> —Don Rittner

Each echo has a set of rules describing the conduct that participants agree to follow. Most rules are developed to ensure the enjoyment and productivity of all participants. For example, the use of profanity and name-calling is discouraged. Infractions of the rules can result in a "flaming"—a sound scolding from the entire membership of that echo. If you are unsure about the rules and etiquette for an echo, you can request them from the moderator or your local SysOp.

Many FidoNet echoes are useful for sharing information and ideas with other environmentalists or for environmental research; a representative sample of those echoes is listed here. Not all Fido boards carry every echo. You can send a request to the SysOp of your local Fido board asking that it carry the echoes in which you are interested. Chances are the SysOp will accommodate you.

For a complete list of all echoes carried on FidoNet, ask your SysOp to send you a copy of the official "EchoList." Mike Fuchs (his Fido ID is 1:1/201) is the keeper of the list. If you don't find an echo that covers your interest, you can create one yourself, provided enough people are interested in participating.

The following list of Fido echoes contains the name of each echo, the moderator, his or her node number, and how the echo is distributed or gatewayed to other networks. Some echos are by invitation only, are for SysOps only, or require moderator approval.

Environment and Science

Environmental Issues
This Canadian echo explores environmentally safer alternatives to materials and substances used in the home and other places. Questions and answers are also fielded.

> Moderator: James Farrow
> Node: 1:163/142
> Distribution: Backbone

Hazardous Waste Management
Discussions pertain to hazardous waste management in general.

> Moderator: Hector Mandel
> Node: 1:233/15
> Restrictions: Need moderator approval to join
> Distribution: 1:233/13

Indian Affairs
Forum topics include current affairs, anthropology, archaeology, and history (including trivia questions). The SysOp welcomes all with an interest in Native American culture, art, and religion.

> Moderator: SysOp Tiwa
> Node: 1:320/999
> Distribution: Backbone

Sustainable Agriculture
This forum is for the discussion and exchange of information by devotees and practitioners of natural, biologically oriented, low-input, regenerative agriculture.

> Moderator: Lawrence London
> Node: 1:151/502
> Distribution: Backbone

The Earth Echo
An echo solely dedicated to environmental issues.

> Moderator: Gordon Gattone
> Node: 1:18/8
> Distribution: Backbone

International Forum, Where Borders Don't Matter
A good echo in which to discuss global environmental issues.

> Moderator: Joaquim Homrighausen
> Node: 1:135/20
> Distribution: Backbone, zone 2, zone 3
> Gateways: Zone 2 via 2:310/11

Evolutionary Mechanism Theory Discussion
For discussion of evolutionary theory, field biology, and observations in ecology, genetics, and population biology. It is specifically geared toward providing a communications channel for scientists organizing against antiscience creationist forces.

> Moderator: Wesley Elsberry
> Node: 8:930/303

Health Physics Society Conference
Industrial hygiene and radiation topics are covered in this echo, which is restricted to radiation safety professionals and serious students. The echo links to the Internet by way of the hps@ux1 Internet mailing list.

> Moderator: Hector Mandel
> Node: 1:233/15
> Restrictions: Need moderator approval to join
> Distribution: 1:233/13
> Gateways: Internet via 1:233/13

Geography

Mapping
This echo is for land surveyors, navigators, cartographers, geodesists, geographers, and other people interested in mapping issues. You could pick up some good information here on land issues.

> Moderator: Maynard Riley
> Node: 1:115/678
> Distribution: 1:115/678, 1:132/125

Health and Safety

Hunger Conference
This echo discusses the social aspects of ending starvation, death by starvation, malnutrition, and other similar social violence. Global warming trends will have a serious impact on food production, and this echo is a good place to discuss the implications on population growth and movements.

> Moderator: Tracy Graves
> Node: 1:159/500
> Distribution: National, international

Labor and Union News Conference
No other large group is affected as much as labor by workplace health and safety issues, toxic waste exposure and so on. This echo contains news articles and analysis of the world labor and union movement.

> Moderator: William Bowles
> Node: 1:107/607
> Distribution: Regional

Radiation Safety

A general discussion of radiation safety issues and current events relative to the nuclear industry, for both professionals and the general public.

> Moderator: Jamie Prowse
> Node: 1:260/160
> Distribution: Backbone

Safety Professional's Forum

A forum in which safety professionals can discuss issues related to industrial hygiene, radiation safety, chemical safety, fire safety, and related topics. This echo is linked to the safety@uvmvm.bitnet mailing list.

> Moderator: Hector Mandel
> Node: 1:233/15
> Restrictions: Need moderator approval to join
> Distribution: 1:233/13
> Gateways: BITNET via 1:233/13

Politics

Controversial

A forum for discussion of controversial topics, current events, and contemporary attitudes, with no (or very few) holds barred, according to the moderator.

> Moderator: Charles Hill
> Node: 1:147/54
> Distribution: Backbone

Defense Issues

An echo geared for news and analysis on the defense industry, disarmament, and economic conversion of defense industries.

This is a good place to discuss the environmental consequences of nuclear waste production and problems such as the health effects of working in nuclear material development.

> Moderator: William Bowles
> Node: 1:107/607
> Distribution: Regional

Networking for Change

How can you use telecommunications and networking for social change and the development of sound national environmental policy? This echo provides ideas and discussions on using networking in all its forms to promote social change.

> Moderator: William Bowles
> Node: 1:107/607
> Distribution: Regional

Politics

An open debate on general politics, including environmental issues. Billed as a "Letters to the Editor" column you can answer.

> Moderator: Edward Cleary
> Node: 1:157/200
> Distribution: Backbone

News

Local National Regional News

This echo is a good alternative news outlet through which users and SysOps can distribute information quickly and easily.

> Moderator: Jake Hargrove
> Node: 1:301/1
> Distribution: National, regional, local

Media Transcription Service
Good source for transcribed news articles and analysis of world news stories, texts of speeches, and interviews with world news makers.

> Moderator: William Bowles
> Node: 1:107/607
> Distribution: Regional

News of the U.S. and the World
This echo features current events and items from today's news that are frequently overlooked in the mass media. You'll find both national and international news stories, with an emphasis on social, environmental, and labor issues. Available from the zone 1 Echomail Backbone and in zone 2 from 2:300/51 and others.

> Moderator: Randy Edwards
> Node: 1:141/552
> Distribution: International, zone 1, Backbone

U.N.—NGOs
This echo features news and discussion of nongovernmental organizations (NGOs) and the U.N. transnational organizations.

> Moderator: William Bowles
> Node: 1:107/607
> Distribution: Regional

United States—South Africa
USA-SA is an international echo that exchanges with the United States, South Africa, Puerto Rico, Great Britain, and a few other places. You can learn more about these countries and communicate with fellow environmentalists.

> Moderator: Steve Richardson
> Node: 1:154/300
> Distribution: Zone 1 Backbone, 1:154/300, 1:367/8, 2:254/350, 5:491/6
> Gateways: Zone 2 via 1:154/300, zone 5 via 1:154/300, Eggnet via 1:367/8

U.S.A.—Europe Link
A nontopical, nonpolitical conference designed to enhance the exchange of information between the United States and European countries. The two moderators offer mail and file transfer options.

> Moderator: Brenda Donovan
> Node: 1:202/701
> Distribution: Backbone, zone 2

Native American NewsMagazine
First Native American magazine with hard-hitting articles from the AP wire service and Native American newspapers around the country. This is a "read-only" echo.

> Moderator: Randy Redhawk
> Node: 1:284/2
> Distribution: Backbone

Native American Controversy
A complement to Native American NewsMagazine, this echo allows users to voice their opinions, write letters to the editor of the magazine, and help promote Native American awareness.

> Moderator: Randy Redhawk
> Node: 1:284/2
> Distribution: Backbone

Recreation

Hiking, Mountain Climbing, and Camping
A good echo for discussion of outdoor activities. Topics include recommended locations, equipment, tips, and related subjects.

> Moderator: Lance Rasmussen
> Node: 1:350/90
> Restrictions: Moderator approval required

Spelunker's Forum
This echo for discussions about caving is linked to the Cavers-Request@m2c.org Internet mailing list.

> Moderator: Hector Mandel
> Node: 1:233/15
> Restrictions: Need moderator approval to join
> Gateways: Internet via 1:233/13

Wilderness Experience
An international forum on wilderness camping, backpacking, hiking, canoeing, mountaineering and related topics. Users range from the occasional weekend hiker to the expedition leader. Most areas of the United States are represented, and a European connection is operational at 2:215/106 in the Netherlands.

> Moderator: John Perkins
> Node: 1:260/315
> Distribution: Zone 1, zone 2
> Gateways: 2:512/106

Technology

Information Power
This echo area is established for librarians exploring technological access to information. This includes, but is not limited to,

online database searching, CD-ROM, cable-access news, interactive media, automated circulation, and online public access systems.

> Moderator: Janet Murray
> Node: 1:105/23
> Distribution: National, backbone

Technology Education
An echo designed to encourage students to learn telecommunications hands-on. Schools are urged to join, especially those in New York State, where computer literacy is now compulsory in primary and secondary education. Students can send messages to "pen pals." Any SysOp willing to let local schools use his or her BBS for this echo is especially welcome. The BBS must be rated "G"—suitable for audiences of all ages.

> Moderator: Doug Purdy
> Node: 1:267/41
> Distribution: New York State and surrounding area

Wilderness is a resource that can shrink but not grow.

Aldo Leopold, 1949

BITNET

BITNET (Because It's Time Network) is one of three interconnected networks that span 38 countries and link up more than 2000 academic and research organizations. The North American link is also called BITNET, the European link is called EARN (European Academic Research Network), and the Canadian link is called NetNorth. All three operate under the umbrella of the Corporation for Research and Educational Networking (CREN).

BITNET's primary use is correspondence among researchers around the world. In addition to BITNET's own nodes, the network has gateways for the exchange of electronic mail with the Internet, FidoNet, and many other networks worldwide.

With a BITNET electronic address, you can send electronic mail or files instantly to researchers in Africa, Asia, Europe, South America, the Middle East, and North America. You can also participate in hundreds of "mailing lists" (discussion groups), many of which are of interest to environmentalists.

BITNET has had an amazing growth. It started when computers at City University of New York and Yale were linked to each other in 1981. Originally, the universities wanted to provide communications to their computer users with no requirements, restrictions, or fees. The network was originally restricted to universities with IBM mainframes. Later, software was developed to allow the network to run on UNIX or Digital Vax computers running the VMS operating system, as well as on other systems that use IBM's RSCS/NJE mail and file transfer protocols.

Computers on BITNET interconnect through leased phone lines, and other permanent links, and the data is quickly transmitted at 9600 bps. BITNET is a "store and forward" network, meaning that information sent from a given BITNET computer or node, is forwarded through intermediate nodes, until it reaches its destination.

How to Get Online

To access BITNET from your own computer, you must set up an account at a facility that is a member of BITNET. Getting a BITNET account is possible if you are a student or faculty member at or otherwise affiliated with a university that is a BITNET node. Also, many universities grant "external" accounts to members of nonprofit organizations or the general public. BITNET has more than 2000 nodes, with about 500 in the United States alone.

Once you find a local node, you must obtain a user ID from that institution. Then, since nodes use different operating systems, you need to learn the text editor of your local node and how to send mail to BITNET.

To learn how to get online or for further information about BITNET, contact the BITNET Network Information Center, EDCUCOM, 112 16th St. NW, Ste. 600, Washington, DC, 20036; 202/872-4200.

Cost

In most cases, the service is free or available at a minimal cost. A small hourly fee is usually charged for external accounts.

Rain Forest Networking: Bringing Scientists and Activists Together

When Aldo de Moor, a fourth-year Information Management student at Tilburg University in the Netherlands, decided to help preserve the rain forest, he set out to do it on a global scale. He formed the Rain Forest Network Bulletin on BITNET.

"Solving complex rain forest problems demands an interdisciplinary scientific approach," Aldo believes. "Many scientists are studying rain forest issues, but a major barrier to developing a real interdisciplinary approach is the lack of exchange among the different scientific disciplines: economics, biology, political science, and so forth. With the Rain Forest Network, I hope to get the best minds of the various disciplines focused on the 'big picture.'"

Aldo feels that by exchanging information, establishing contacts, and motivating each other, scientists and activists can work together to find sustainable rain forest management methods.

Computer networking is still in its infancy in Europe, Aldo explains. "In the Netherlands, some environmental organizations are already using GreenNet, but I am trying to convince them that global networking can be a very powerful tool."

One of the major goals for the Rain Forest Network Bulletin is for it to become a scientific think tank for evaluating action plans from nongovernmental agencies, governments, and private organizations—to be a creative forum for new ideas.

Aldo is especially excited about the possibility of live teleconferencing on BITNET. "People doing research on the same subject in Brazil, the United States, and Europe could have a 'telemeeting,' without any traveling. In this way, people from the rain forest countries, like Brazil, can actively participate in making the decisions that directly affect their homeland," he says.

> Computer networking brings people together, including concerned scientists who don't have the time or dollars to spend traveling around the world sharing their expertise with environmental activists. "Too often, these two worlds work apart from each other, which is a pity; science can offer the theories, environmental movements the practice. Working together, people can do so much more!"
>
> The spirited exchanges taking place on the Rain Forest Network are not limited to the scientific and the political. Aldo encourages participants to examine what he feels are the complex social roots of rain forest issues: hypocrisy, egotism, lack of respect, inversion of values, and shortsightedness.
>
> Although the network has been around for only a short time, it has already linked hundreds of scientists, students, professors, and interested citizens from around the world.
>
> Aldo's BITNET address is ademoor@KUB.NL.
>
> —Don Rittner

Electronic Mail

Addressing Mail

All computers on BITNET can use Internet-style domain addresses; the format for this type of address is username@subdomain. "Username" refers to the ID a person receives when he or she establishes an account with BITNET. The "subdomain" is the ID of the host site at which that user receives mail. For example, if you wanted to send me mail on BITNET, you would address it to

```
drittner@albnyvms
```

`Drittner` is the user ID assigned to me. `Albnyvms` is the host site, a BITNET node located at the State University of New York in Albany, New York.

I should warn you that all is not rosy in the network world. With more than a hundred networks, many with their own operating systems and protocols, routing mail from one network to another can get confusing. While many systems use gateways to forward mail automatically, sometimes you need to provide help by adding more "path" information in your destination address—that is, by specifying what route the mail needs to take to get to its recipient. *A Directory of Electronic Mail: Addressing and Networks* by Donnalyn Frey and Rick Adam explains in detail the addressing schemes that all the networks use (see Appendix E).

If you know that someone is on the network at the same time you are you can send that person *interactive* ("real-time") mail. It takes only a few simple commands, although those vary depending on the host computer's software and operating system.

You learn the commands in various ways: from a user manual, from the host, or by trial and error. As an example, someone on an IBM mainframe using the VM/CMS system could send me a message by typing:

`TELL drittner@albnyvms Please call me tomorrow.`

Since I'm on the VAX/VMS system, I would use slightly different commands to respond:

`SEND userid@node "OK, will call you."`

Gateways to Other Networks

BITNET can exchange mail with FidoNet, Internet, and other networks through gateways that direct your mail to the proper destination. At the present time, BITNET (including EARN and NetNorth) has gateways to the following networks:

- FidoNet (worldwide)
- Internet (NSFNET, CSNET, and others)

- HEPnet (high energy physics network)
- MFEnet (magnetic fusion energy network)
- IBM-VNET (IBM corporation network)
- CDNnet (Canadian research and education network)
- DFNnet (West German research network)
- HEAnet (Irish higher education authority network)
- INFnet (Italian nuclear physics network)
- JANET (United Kingdom joint academic network)

Sending Files

Sending files is very similar to sending interactive messages. For example, I would send someone a file, using the VAX/VMS system, like this:

List Servers

BITNET uses *servers* to store and forward information and files. Each server resides on a host computer and has an ID similar to a user ID. The server responds to commands you send to retrieve documents, help files, software, mailing lists, and other data.

BITNET has many servers, each with its own contents and purpose. *Listservers* control mailing lists, and *Fileservers* control file lists, although some Listservers control both. Other servers specialize in subjects ranging from the biological sciences to poetry. To send for a list of BITNET servers, you can use the interactive command, `SEND listserv@bitnic GET BITNET SERVERS`. You can also send this request as an e-mail file: If all goes well, the file will be waiting in your mailbox within minutes.

A helpful thing to do when you first log on to BITNET is send HELP (by typing the word HELP) to a server that interests you; it will send you a list of all the commands unique to that server. You can also request from that server a list of the files that reside on it.

BITNET Mailing Lists for Environmentalists

As on FidoNet, interesting discussions go on among users of BITNET. BITNET discussion groups are called "mailing lists" and include forums, digests, and electronic magazines. About 900 discussion groups can be found on BITNET. Topics pertinent to environmental study and networking include agriculture, archaeology, astronomy, astrophysics, biology, chemistry, current events, news, ecology, the environment, economics, education, energy, genetics, geology, geography, government agencies, higher education, history, medicine, natural resources, physics, space sciences, and world politics.

To join a mailing list, you need to subscribe. The format for doing so is SEND *node* sub *name of mailing your name*. For example, to subscribe to the BioSphere Newsletter, an environmental newsletter from the State University of New York at Buffalo, I would send the following command:

`SEND listserv@ubvm sub biosph-l Don Rittner`

Provided the mailing list were open to all, I would receive a reply telling me I was accepted, and I would thenceforth automatically get mail addressed to the biosph-l address.

The members of forums correspond by sending mail to a specified address, where their mail is automatically forwarded to all forum members. Digests are organized and edited by moderators, who periodically send out member contributions to the people on the mailing list. Electronic magazines contain features and columns like their printed counterparts and are also distributed periodically to subscribers.

Table 4-1 lists some of the mailing lists to which you can subscribe on BITNET.

TABLE 4-1 BITNET Mailing Lists	Networkwide ID	Full Address	List Title
	ECONET	ECONET@MIAMIU	(Peered) A discussion of Ecology and Environment
	ADMIN-L	ADMIN-L@ALBNYDH2	New York State Department Of Health
	ADMINISTRATIVE-INFORADMRA-L	ADMRA-L@ALBNYDH2	Adirondack Medical Records Association List
	AG-EXP-L	AG-EXP-L@NDSUVM1	AG-EXP-L Ag Expert Systems
	AG-FORST	AG-FORST@IRLEARN	BIOSCI AgroForestry Bulletin Board
	AGFTECH	AGFTECH@DEARN	AGF-Subnetz-Koordinatoren
	AGPGMR-L	AGPGMR-L@WSUVM1	Ag Programmers Inter/Intra-Disciplinary
	AGRIC-L	AGRIC-L@UGA	Agriculture Discussion
	ALTANRCNET	NRCNET-L@UALTAVM	National Research Council Network
	ALTERNAT	ALTERNAT@NDSUVM1	Alternatives Journal
	ANTHRO-L	MD45@CMUCCVMA	Society for Scientific Anthropology
		ANTHRO-L@UBVM	General Anthropology Bulletin Board
	AOBULL	AOBULL@ALBNYDH2	New York State Department of Health Area Office
	AOBULL-L	AOBULL-L@ALBNYDH2	New York State Department of Health Area Office
	APNET-L	APNET-L@JPNSUT00	Asia Pacific Network
	AQUIFER	AQUIFER@IBACSATA	Pollution and Ground Water
	BEE-L	BEE-L@ALBNYVM1	Discussion of Bee Biology
	BIO-CONV	BIO-CONV@IRLEARN BIOSCI	Bioconversion Bulletin Board
	BIO-JRNL	BIO-JRNL@IRLEARN BIOSCI	Bio-Journals Bulletin Board
	BIO-NAUT	BIO-NAUT@IRLEARN BIOSCI	Bionauts Bulletin Board
	BIO-SOFT	BIO-SOFT@IRLEARN BIOSCI	Software Bulletin Board
	BIOJOBS	BIOJOBS@IRLEARN BIOSCI	Employment Bulletin Board
	BIOMATRX	BIOMATRX@IRLEARN BIOSCI	Bio-Matrix Bulletin Board
	BIOMET-L	BIOMET-L@ALBNYDH2	Bureau Of Biometrics at ALBNYDH2

Networkwide ID	Full Address	List Title
BIONEWS	BIONEWS@IRLEARN BIOSCI	BioNews Bulletin Board
BIOSPH-L	BIOSPH-L@UBVM	Biosphere, Ecology, Discussion List
BIOTECH	BIOTECH@IRLEARN BIOSCI	BioTech Bulletin Board
BIOVOTE	BIOVOTE@IRLEARN BIOSCI	Ballot Box
BIRD_RBA	BIRD_RBA@ARIZVM1	National Birding Hotline Cooperative
BRINE-L	BRINE-L@UGA	Brine Shrimp Discussion List
CCMEDH-L	CCMEDH-L@TAMVM1	Cross Cultural Health and Medical List
CHEM-L	CHEM-L@UOGUELPH	Chemistry Discussion
CHEME-L	CHEME-L@PSUVM	Chemical Engineering List
CHEMED-L	CHEMED-L@UWF	Chemistry Education Discussion List
CHEMIC-L	CHEMIC-L@TAUNIVM	Chemistry in Israel List
CIVIL-L	CIVIL-L@UNBVM1	Civil Engineering Research & Education
CLADE	CLADE@MSU	Phylogenetics
CLAN	CLAN@FRMOP11	Cancer Liaison and Action Network
CLASS-L	CLASS-L@SBCCVM	Classification, Clustering, and Phylogeny
CLIMLIST	CLIMLIST@OHSTVMA CLIMLIST	List of Climatologists
CONSLINK	CONSLINK@SIVM	Discussion on Biological Conservation
EBCBBUL	EBCBBUL@HDETUD1	Computers in Biotechnology, Research and Education
EBCBCAT	EBCBCAT@HDETUD1	Catalogue of "Biotechnological" Software
ECONOMY	ECONOMY@TECMTYVM	Economic Problems in Less Developed Countries
EDPOLYAN	EDPOLYAN@ASUACAD	Professionals and Students Discussing Education
EMFLDS-L	EMFLDS-L@UBVM	Electromagnetics in Medicine, Science & Communities
ENERGY-L	ENERGY-L@TAUNIVM	Energy List
ENVBEH-L	ENVBEH-L@POLYGRAF	Forum on Environment and Human Behavior
EPID-L	EPID-L@QUCDN	Topics in Epidemiology and Biostatistics
FORUMBIO	FORUMBIO@BNANDP11	Forum on Molecular Biology
GENBANKB	GENBANKB@IRLEARN BIOSCI	GENBANK-BB Bulletin Board

Networkwide ID	Full Address	List Title
GENE-EXP	GENE-EXP@IRLEARN BIOSCI	Gene-Expression Bulletin Board
GENE-ORG	GENE-ORG@IRLEARN BIOSCI	Genomic-Organization Bulletin Board
GEODESIC	GEODESIC@UBVM	List for the Discussion of Buckminster Fuller
GEOGRAPH	GEOGRAPH@FINHUTC	Geography
GNOME-PR	GNOME-PR@IRLEARN BIOSCI	Human Genome Program Bulletin Board
GOVDOC-L	GOVDOC-L@PSUVM	Discussion of Government Document Issues
HEALTH-L	HEALTH-L@IRLEARN	International Discussion on Health Research
HEALTHCO	HEALTHCO@RPICICGE	Communication in Health/Medical Context
	HEALTHCO@RPIECS	Communication in Health/Medical Context
HGML-L	HGML-L@YALEVM	Human Gene Mapping Library
HORT-L	HORT-L@VTVM1	VA Tech Horticulture Dept.— Monthly Releases
ITRDBFOR	ITRDBFOR@ASUACAD	Dendrochronology Forum
MEDGEN-L	MEDGEN-L@INDYCMS	Medical Genetics List
MEDIA-L	MEDIA-L@BINGVMB	Media in Education
MEDNEWS	MEDNEWS@ASUACAD	MEDNEWS—Health Info-Com Network Newsletter
MMICRONET	MICRONET@UOGUELPH	Fungus and Root Interaction Discussion
NEWS-L	NEWS-L@ALBNYDH2	New York State Department Of Health News
NEWSB-L	NEWSB-L@BUACCA	College News Bureaus List
NEWSE-D	NEWSE-D@INDYCMS	News of the Earth Distribution
NEWSE-L	NEWSE-L@INDYCMS	News of the Earth Letters
NEWSE-S	NEWSE-S@INDYCMS	News of the Earth Supplements
ODP-L	ODP-L@TAMVM1	Ocean Drilling Program Open Discussion List
PHOTOSYN	PHOTOSYN@TAUNIVM	Photosynthesis Researchers' List
PHOTREAC	PHOTREAC@JPNTOHOK	Electro- and Photo-Nuclear Reaction Discussion
PHYS-L	PHYS-L@UCF1VM	Forum for Physics Teachers
PHYS-STU	PHYS-STU@UWF	Physics Student Discussion List
PIR-BB	PIR-BB@IRLEARN BIOSCI	PIR Bulletin Board
POLCOMM	POLCOMM@RPICICGE	Political Communication
	POLCOMM@RPIECS	Political Communication
POLI-SCI	POLI-SCI@RUTVM1	Political Science Digest

Networkwide ID	Full Address	List Title
POLITICS	POLITICS@UCF1VM	Forum for the Discussion of Politics
POP-BIO	POP-BIO@IRLEARN BIOSCI	Population Biology Bulletin Board
RECYCLE	RECYCLE@UMAB	Recycling in Practice
RESEARCH	RESEARCH@IRLEARN BIOSCI	Research-News Bulletin Board
SCI-RES	SCI-RES@IRLEARN BIOSCI	Science-Resources Bulletin Board
SEISM-L	SEISM-L@BINGVMB	Seismological Data Distribution
SEISMD-L	SEISMD-L@BINGVMB	Seismological Discussion
SFER-L	SFER-L@UCF1VM	South Florida Environmental Reader
STATEPOL	STATEPOL@UMAB	Politics in the American States
SUNSPOTS	SUNSPOTS@FRULM11	(Peered) Sun Spots Discussion
TAXACOMA	TAXACOMA@MSU	Taxacom Announcements
TAXACOMR	TAXACOMR@MSU	Taxacom Request
TAXACOMT	TAXACOMT@MSU	Taxacom Technical
VOLCANO	VOLCANO@ASUACAD	VOLCANO
WX-PCPN	WX-PCPN@UIUCVMD	WX-PCPN Precipitation WX Products
WX-SPOT	WX-SPOT@UIUCVMD	WX-SPOT Weather Spotters and Coordinators
WX-STLT	WX-STLT@UIUCVMD	WX-STLT Satellite Interpretive Messages
WX-SWO	WX-SWO@UIUCVMD	WX-SWO Severe Weather Outlooks
WX-TALK	WX-TALK@UIUCVMD	WX-TALK General Weather Discussions and Talk
WX-TROPL	WX-TROPL@UIUCVMD	WX-TROPL Tropical Storm and Hurricane
WX-WATCH	WX-WATCH@UIUCVMD	WX-WATCH WX Watches and Cancellations
WX-WSTAT	WX-WSTAT@UIUCVMD	WX-WSTAT WX Watch Status and Storm Reports
WXSPOT-L	WXSPOT-L@UIUCVMD	WXSPOT-L Weather Spotters and Coordinators

CHAPTER 5

The realization that our small planet is only one of many worlds gives mankind the perspective it needs to realize sooner that our own world belongs to all of its creatures.

Arthur C. Clark

Usenet

Usenet, or User's Network, is a public series of conferences—called *newsgroups*—carried primarily over UNIX-based computers. Many Usenet conferences are of use to environmental researchers.

Usenet began in 1979 on two hosts and is now distributed to over 250,000 readers on about 18,000 hosts spanning five continents. Usenet participants access the network on all kinds of equipment, from small PCs at home to supercomputers at large universities, research organizations, and even corporations. Although Usenet newsgroups are primarily distributed over computers that use the UNIX operating system, Usenet has gateways to Internet, BITNET, and other networks.

Usenet participants subscribe to newsgroups that interest them. Subscribers post messages—called "articles"—and replies all of which are circulated to all other members of that particular newsgroup. Unlike FidoNet and BITNET, Usenet itself does not support e-mail. Another network, UUCP, carries the e-mail generated for Usenet and, in fact, transports the Usenet transmissions themselves from host to host.

How to Get Online

Usenet is similar to FidoNet in that there is no central administration and anyone can join. To participate, you must first get an account with a network or institution that carries Usenet newsgroups. Places to find Usenet newsgroups include:

Your local college or university
If you are a student and your school is a Usenet site, you can probably get an account and participate for no cost. If you are not a student, check with your local academic institution; many grant external accounts to qualified organizations and individuals.

The WELL
The WELL online service, described in chapter 12, carries many of the Usenet newsgroups. The WELL is a subscription service that carries a monthly membership fee and hourly connect charges.

EcoNet
Described in chapter 10, EcoNet carries most of the Usenet newsgroups. EcoNet is also a subscription service.

FidoNet
A number of the FidoNet boards described in chapter 3 carry Usenet newsgroups.

UUNET
If you have trouble finding a local Usenet feed, contact UUNET Communications Services (see Appendix C) and subscribe to UUCP, the network that carries the Usenet newsgroups from host to host. UUNET will set up your personal computer as a "UNIX site" so that you can subscribe to the Usenet newsgroups that interest you.

Other personal bulletin board services
BBSs across the country carry Usenet newsgroups. Phil Eschallier, a Usenet participant, maintains an updated list of these bul-

letin boards called nixpub (for UNIX Publication). For a copy of the list, contact Phil at email address: phil@ls.com (or on UUCP at !uunet!gn p1!phil), or on CompuServe at 71076,1576.

Once you find a host site or a BBS that carries Usenet newsgroups, check to see whether it carries any in your area of interest. (Since Usenet carries some 4 megabytes of information per day, it is not possible for each host site to carry all newsgroups.) If the site does not carry the newsgroups that interest you, ask the SysOp or the host of the site to consider carrying them.

Software programs called *readers* allow you to read and reply to Usenet articles. Three common readers are vnews and rn (which have screen-oriented interfaces) and readnews (which is line oriented). These programs are found on most Usenet sites, along with instructions on how to use them (see Figure 5-1). A new commercial program called µAccess from ICE Engineering (8840 Main St., Whitmore Lake, MI 48189), gives Macintosh users access to Usenet newsgroups using the familiar Mac interface.

Cost

The costs of the network are shared by participating hosts, and each host pays for its own transmission costs. Usenet is a noncommercial network, but some private bulletin boards and on-line services charge for access to their Usenet areas.

Usenet Newsgroups

The over 500 Usenet newsgroups cover a variety of subjects, many of interest to those doing environmental research and networking. The major Usenet newsgroups are distributed worldwide, and fall into seven main categories, as shown in Table 5-1.

Newsgroup names consist of an abbreviation of the category name (see Table 5-1) followed by a period, and an abbreviation of the group name. For example, the name *sci.bio* signifies that this group—biology—falls in the science category. Subgroups

FIGURE 5-1

```
                        White Knight 11.08
 00:28:09              Serial Port Settings  Pause    ^S   ^Q   ^C
                       2400-E-7-1-FULL       Remote

> energy sources. Similarly, there is no reason why irrigation must
> use fossil water. We do so now because it's cheaper.
>

The use of ground (fossil) water in indescriminate quantities is yet another
example of the flagrant misuse of a resource just because it is cheap. As
with oil and other fossil fuels, the charge to the consumer should, in some
way, account for the cost of replacing that resource. In the case of ground
water the user fee might be increased. Any such increase must be invested in
the development of a replacement water source.

By charging more for the resource we encourage conservation. By reinvesting
the proceeds we ensure the future of the resource.

-------------------------------------------------------------------------
| joe houghtaling          | "In my opinion television | computer science dept. |
| jh0576@leah.albany.edu   |  validates existence."    | suny at albany         |
|                          |             -calvin       | albany, ny 12222 usa   |
"/tmp/posta08145" 29 lines, 1662 characters

What now? [send, edit, list, quit, write, append] s
Posting article...
Article posted successfully.
[Hit return to continue]_

CMD:
```

A typical article posted on Usenet using the vnews reader. This is a response to a previously posted article.

are designated by an additional period and abbreviation; for example, *sci.math.stat* stands for the science/mathematics/statistics newsgroup.

Most newsgroups distribute everything sent in by subscribers. Others are *moderated*, meaning a moderator screens submissions to determine which get posted into the newsgroup.

Newsgroups are created and deleted democratically by votes in a newsgroup called news.groups. Anyone can subscribe to this newsgroup.

TABLE 5-1
Categories of Usenet Newsgroups

Category	Newsgroup Prefix
Computers	comp
Recreation	rec
Science	sci
Social issues	soc
About Usenet groups	news
Chatter	talk
Miscellaneous	misc

Table 5-2 lists Usenet newsgroups worldwide that are of interest to environmentalists. You can find most of the groups on any site that carries Usenet newsgroups.

TABLE 5-2 Usenet Newsgroups

Newsgroup	Description
comp.risks	Risks to the public from computers (moderated)
comp.society	The impact of technology on society (moderated)
sci.aquaria	Scientifically oriented postings about aquaria
sci.astro	Astronomy discussions and information
sci.bio	Biology and related sciences
sci.chem	Chemistry and related sciences
sci.econ	Economics
sci.edu	Science education
sci.energy	Discussions about energy, science, and technology
sci.environment	Discussions about the environment and ecology
sci.med.physics	Issues of physics in medical testing and care
sci.military	Discussion about science and the military (moderated)
sci.misc	Short-lived discussions on subjects in the sciences
sci.research	Research methods, funding, ethics, etc.
sci.skeptic	Skeptics discussing psuedoscience
sci.space	Space, space programs, space-related research
sci.virtual-worlds	Modeling the universe (moderated)
soc.culture.arabic	Technological and cultural issues
talk.origins	Evolution versus creationism (sometimes hot!)

A number of Usenet newsgroups have Internet equivalents that are shared by the two networks, expanding participation on the topic. Table 5-3 lists ones of interest to environmentalists.

TABLE 5-3 Internet/Usenet Equivalents

Internet List	Usenet Newsgroup
physics@unix.sri.com	sci.physics
space@andrew.cmu.edu	sci.space

Alternative Newsgroups

A number of newsgroups outside the seven main categories are distributed as "alternative" newgroups.[1] Alternative newsgroups that are useful for environmental research include the following alternative groups (a subset of alternative newsgroups): BioNet, ClariNet, and Inet/ddn.

Alternative Groups

The alternative groups (*alt*) are a small collection of newsgroups distributed by sites that choose to carry them. You can join the "alt subnet" by finding a site in your area that carries the groups. Either send mail to the administrators of the sites you connect to, or post something to a local "general" or "wanted" newsgroup in your area. If no sites nearby distribute the alternative groups, you can get them from UUNET. Representative groups are listed in Table 5-4.

TABLE 5-4 Alternative Groups	Group	Description
	alt.activism	Activities for activists
	alt.great-lakes	Discussions of the Great Lakes and adjacent regions

Bionet

Bionet is a collection of newsgroups of interest to biologists and is carried by a growing number of hosts including rutgers, phri, mit-eddie, ukma, and all of the machines at UCSD (see Table 5-5). Contact Eliot Lear (Usenet@NET.BIO.NET) for more details. Most of these groups have an equivalent on BITNET.

---

[1]Special thanks to Gene Spafford for a list of these.

TABLE 5-5 Bionet Newsgroups

Bionet Name (BITNET Name)	Description
bionet.agroforestry (AG-FORST)	Discussion of Agroforestry
bionet.general (BIONEWS)	General BIONET announcements
bionet.jobs (BIOJOBS)	Job opportunities in science
bionet.journals.contents (BIO-JRNL)	Contents of biology journal publications
bionet.molbio.bio-matrix (BIOMATRX)	Computer applications to biological databases
bionet.molbio.embldatabank (EMBL-DB)	Information about the EMBL nucleic acid database
bionet.molbio.evolution (MOL-EVOL)	How genes and proteins have evolved
bionet.molbio.gene-org (GENE-ORG)	How genes are organized on chromosomes
bionet.molbio.genome-program (GNOME-PR)	Human Genome Project issues
bionet.molbio.methds-reagnts (METHODS)	Requests for information and lab requests
bionet.molbio.news (RESEARCH)	Research news of interest to the community
bionet.molbio.swiss-prot (SWISSPRT)	Discussion on the SWISS-PROT database
bionet.population-bio (POP-BIO)	Technical discussions about population biology
bionet.sci-resources (SCI-RES)	Information about funding agencies, etc.
bionet.software (BIO-SOFT)	Information about software for biology
bionet.technology.conversion (BIO-CONV)	Technology to convert waste and biomass
bionet.users.addresses	Who's who in biology

ClariNet

ClariNet consists of newsgroups gatewayed from commercial news services and other "official" sources. More information may be obtained by sending mail to info@clarinet.com. Table 5-6 lists some ClariNet newsgroups.

TABLE 5-6 ClariNet Newsgroups	Group	Description
	clari.news	ClariNet UPI general news wiregroups
	clari.tw	ClariNet UPI technology-related news wiregroups
	clari.nb	ClariNet Newsbytes Information Service newsgroups
	clari.tw.computers	Computer industry, applications, and developments
	clari.tw.environment	Environmental news, hazardous waste, forests (moderated)
	clari.tw.misc	General technical industry stories (moderated)
	clari.tw.nuclear	Nuclear power and waste (moderated)
	clari.tw.science	General science stories (moderated)
	clari.tw.telecom	Phones, satellites, media and general telecommunications (moderated)
	clari.nb.telecom	Newsbytes telecommunications and online industry news (moderated)
	clari.news.interest.animals	Animals in the news (moderated)

Inet/ddn

The *inet/ddn* distribution consists of newsgroups bearing names similar to traditional Usenet groups but corresponding to Internet discussion lists. The groups are circulated by means of the NNTP transport mechanism among sites on the Internet in an attempt to reduce the number of copies of these groups flowing through the mail. Further details may be obtained by writing to

Erik Fair at fair@ucbarpa.berkeley.edu. Table 5-7 lists a few pertinent inet/ddn groups.

TABLE 5-7 Inet/ddn Newsgroups	Group	Description
	comp.ai.edu	Applications of artificial intelligence to education
	comp.ai.vision	Artificial intelligence vision research (moderated)
	sci.bio.technology	Any topic relating to biotechnology

UUCP: The Transport System for Usenet

Usenet does not have its own system for transporting newsgroups and e-mail among host sites. Most e-mail is sent via UUCP (UNIX to UNIX Copy Program) software, released in 1978 by AT&T. UUCP runs primarily on computers that use the UNIX operating system and is estimated to be operating on over 10,000 hosts with a million users.

UUCP is a transport protocol for data communications and a set of files and commands used for communications. You can send mail and files and execute commands on a remote system using UUCP. All you need are the UUCP programs, a modem and a phone line, direct dial, and a public data network. (See Appendix C to find out how you can subscribe to UUNET, a network that provides UUCP.) As with Usenet, each host on UUCP pays for its own calls, and there is no central authority.

Sending mail on UUCP can be confusing. Since UUCP uses "source routing," you need to specify the entire path the mail must follow to get to its recipient. Here's where directories such as Frey and Adam's (see Appendix E) come in handy. As an example, to send me a letter through UUCP, you would have to specify

```
host1!host2!host3!host!drittner
```

These UUCP addressing schemes are being phased out in favor of Internet-style addresses. If you are lucky, your host site

Electronic Newsletter Distribution

Lelani Arris of Alberta, Canada, uploads the proceedings of the Intergovernmental Negotiating Committee for the Framework Convention on Climate Change, held in Nairobi, so these transcripts are available worldwide on the day of the proceedings.

"We get it out all over the world within a few hours of its being published on paper in Nairobi," Arris said. "People are using the information, too; someone in Washington, D.C. is looking at the procedures going on in this Framework Convention, a woman on Pegasus (an Australian environmental network) is rebroadcasting part of it on public radio stations around Australia. Also, environmental journalists appreciate being able to get this information quickly. It's typically not available through United Press International or the Associated Press."

The Framework Convention is a forum for countries worldwide to effectively come up with an agreement that will set limits on the production of greenhouse gases. The NGOs' (Non-Governmental Environmental Organizations') role is to lobby for specific production targets. Readers of the proceedings can easily follow the various positions of the world's governments.

Arris explains, "The electronic newsletters contain developments at the convention: which countries have proposed what commitments to reduce greenhouse gases, who's holding back. They include profiles of NGOs involved in the procedure, and editorial comment. They are only published while these conferences are going on, so there's a quick hit of eight or nine issues over that many days. Then the process starts over two or three months later."

> In 1990, Arris worked out an agreement with the publishers of the newsletters covering the Intergovernmental Negotiating Committee, and began posting them on Econet.
>
> "They started sending me text files of the newsletter from Geneva. I cleaned up the files a little and uploaded them." Arris said. She also sent text by e-mail to users of other networks all over the world, and posted text files in one of the UseNet newsgroups. Arris received many messages from appreciative e-mailers around the world.
>
> In past years, the newsletters would have been available only on paper, and anyone wanting to follow the proceedings of such a conference would either have to be there in person, or else would have to wait until copies of the newsletters arrived by mail.
>
> Arris may be found in the User Directory of EcoNet.
>
> —Wendy Monroe

will already have done this, making sending mail as easy as sending mail on BITNET or the Internet. Many networks use UUCP to send e-mail and Usenet newsgroups.

As with BITNET and the Internet, you can subscribe to UUCP-based mailing lists. The following is a partial list of mailing lists available primarily on the UUCP network. Access to some of the lists is restricted; contact list administrators for more information. Whether you are allowed to join those groups is decided by the moderator and the readers.

Animal-rights
An unmoderated list for the discussion of animal rights. Consumers and researchers alike are facing new questions concerning the human animal's treatment of the rest of the animal kingdom. The purpose of this list is to provide students, researchers, and activists a forum for discussing issues such as animal liberation;

consumer product testing; cruelty-free products; vivisection and dissection; medical testing; animals in laboratories; research using animals; hunting, trapping and fishing; animals in entertainment; factory farming; fur; ecology; environmental protection; vegetarianism; vegan lifestyles; and Christian perspectives.

> Contact: Chip Roberson
> Address: animal-rights-request@cs.odu.edu

Beebox

Discussion of beekeeping as a hobby—for the keeper's own consumption and minor sale—emphasizing how to get started, equipment, basic care, control of common and crucial diseases and pests, small-volume low-overhead production, bees for gardening, and references. Also technical apiculture and entomology, and neat recipes for honey.

> Contact: Ken Leonard
> Address: kleonard@gvlv2.gvl.unisys.com or
> ...!burdvax!gvl!gvlv2!kleonard

Cavers

Information resource and forum for anyone interested in exploring caves. To join, send a note to the following address including your geographical location as well as your e-mail address, details of your caving experience, locations where you've caved, your NSS number if you have one, and any other information that might be useful.

> Contact: John D. Sutter
> Address: cavers-request@m2c.org or
> harvard!m2c!cavers-request

Chem-talk

Chemistry teachers and researchers often find that the demands of the profession lead to a reduction in the ability to keep abreast of new data and changes in theories. Sometimes conversation helps to clarify articles, illuminate new perceptions of theories, and sustain scientists on their precarious journey. The creation of an efficient communications network in the form of this mailing list can encourage the dialogue among chemists that is essential in stimulating new ideas.

> Contact: Manus Monroe
> Address: ...!{ames,cbosgd}!pacbell!unicom!manus

Climbing

The climber's mailing list is a forum on all sorts of topics in climbing, from ethics to equipment, from ice climbing to rock climbing to mountaineering.

> Contact: Fritz Nordby
> Address: {ames,rutgers}!cit-vax!climbing-request or
> climbing-request@csvax.caltech.edu

Ethology

An unmoderated mailing list for the discussion of animal behavior and behavioral ecology. Possible topics are new or controversial theories, new research methods, and equipment. Announcements of books, papers, conferences, and new software for behavioral analysis are also encouraged.

> Contact: Jarmo Saarikko
> Address: saarikko@cc.helsinki.fi

Herpetology

The herpetological mailing list promotes the exchange of ideas and information relating to the study of reptiles and amphibians. Discussions include husbandry and veterinary care of captives, field biology and observations, and captive propagation and breeding techniques. The list also provides a forum for the legal exchange of captive specimens. The list stresses conservation and preservation of native wildlife populations.

Contact: Stan Voket
Address: uunet!stpstn!gaboon!herpetology-request or
philabs!crpmks!gaboon!herpetology-request

Killifish

This list is for people who keep or are interested in killifish (family Cyprinodontidae).

Address: killie-request@mejac.palo-alto.ca.us

NativeNet

To provide information about and to discuss indigenous people around the world and current threats to their cultures and habitats (for example, to rain forests).

Contact: Gary S. Trujillo
Address: gst@gnosys.svle.ma.us

Newlists

This is a clearing house for new mailing lists. Subscribers get announcements of new lists that are mailed to this list.

Contact: Marty Hoag
Address: info@vm1.nodak.edu

NewsCom

The purpose of this list, whose full name is Newscommando, is to make available synergies discerned in, and created from, print news media of the last 12 years. Many "facts," particulary scientific ones, have a habit of changing with time. Newscommando shows extreme prejudice toward those articles whose contents exhibit "legs." The depth of insight possible using the information mosaic method can be staggering. A form of electronic magazine, Newscommando can serve as a reference tool, offers unique jump-off points for Medline, PaperChase, and other searches, and in many ways is the "poor man's IdeaFisher/Idea-

Bank." Use "NewsCom request" in the "Subject:" field of message headers to request that NewsCom be sent to your mailbox. Indicate article titles desired or "all" in body of message.

> Contact: Lance Sanders
> Address: starkid@ddsw1.mcs.com

Ocean policy

Discussions of the legal, economic, and military aspects of ocean-use policies. Specific issues include the "law of the sea," pending treaties, the economic implications of EEZs, and the military use of the sea. Coordinated by Dr. Scott Allen, Associate Director, International Law of the Sea Institute, University of Hawaii.

> Contact: Scott Allen
> Address: {dual,vortex}!islenet!scott

SFER-1

SFER-1 (the South Florida Environmental Reader) is a monthly digest of environmental articles of interest to South Florida residents. The newsletter is available in both paper and electronic formats.

> Contact: A. E. Mossberg
> Address: sfer-request@mthvax.cs.miami.edu

Space-activists

The space-activists digest is a public, moderated mailing list for people interested in receiving timely information related to space activism—that is, reports on government activities in this area.

> Contact: Christopher Welty
> Address: space-activists-request@cs.rpi.edu

Wildnet
This list concerns computing and statistics in fisheries and wildlife biology.

> Contact: Eric Woodsworth
> Address: dvinci!ejw or
> woodsworth@sask.bitnet

Is civilization progress? The challenge, I think, is clear; and, as clearly, the final answer will be given not by our amassing of knowledge, or by the discoveries of our science, or by the speed of our aircraft, but by the effect our civilized activities as a whole have upon the quality of our planet's life—the life of plants and animals as well as that of men.

Charles A. Lindbergh, Jr.

CHAPTER 6

Internet

With its broad reach and usage, the federally funded Internet is the closest thing ever developed to a "universal network" linking individuals and organizations around the world. Rather than being a single large network, the Internet is a conglomeration of more than 2,200 smaller networks, connecting more than 50,000 hosts, and an estimated 3 million users. The Internet reaches over 40 countries.

This vast "network of networks" was born in 1969 when the Department of Defense formed a network for the Advanced Research Projects Agency—ARPANET—designed to help government scientists communicate and share information. In 1982, ARPANET split into ARPANET and MILNET, the Military Network. As other networks developed, they became linked by agreeing to use common communications protocols called TCP/IP (Transmission Control Protocol/Internet Protocol).

The National Science Foundation joined Internet in 1986 when the foundation created NSFNET to link several national supercomputer centers to support scholarly research. The NSFNET backbone of the Internet is now composed of 17 networks, connecting to 23 midlevel wide-area networks across the continent. In turn, the midlevel networks link computers in more than 1000 university, government, and commercial research organizations throughout the world. Environmentalists' interest in the Internet lies in NSFNET and its affiliate networks.

NSFNET—the nonmilitary portion of Internet—and MILNET (ARPANET recently dismantled) are now the backbone for the Internet, carrying the burden of the traffic on very fast 56 Kbps (kilobits-per-second) lines or T1 (1.5 Mbps or megabytes-per-second) transmission lines. NSFNET is now increasing the speed of the network to 45 Mbps, making it the world's fastest publicly available network for research and education.

Three primary protocols drive the Internet:

- mail (SMTP)
- file transfer (ftp)
- remote log on (telnet)

This chapter examines each of those functions, as well as Internet resources that you can take advantage of: mailing lists, online libraries, scientific databases, and access to other environmental researchers.

How to Get Online

The Internet, funded by an annual federal subsidy, is intended to support government and academic research. However, those not involved in such research can connect to the Internet in two ways:

- by establishing an account with a research or educational institution that connects to the Internet
- by subscribing to one of the network gateways described in Appendix C

Remember that the Internet is not one network but rather a group of interconnected networks. If you gain access to the Internet, you still may not be able to connect with all of it since many areas are used for private or government research. Other parts allow public access only if you can get the proper clearance.

The best way to get involved is to get access to Internet mail by one of the two methods previously mentioned. Internet mail, which gives you access to the mailing lists described later in the chapter, is in itself a tremendous resource, linking you to many leading research and academic institutions, as well as individuals, worldwide. Take some time to explore areas of the Internet that allow public access. Once you get accustomed to the general organization of the Internet, you may discover additional areas that interest you and find out how to gain access to some of the restricted areas.

Since Internet is command driven and every host requires different commands, this chaper does not attempt to teach you how to use Internet. You will learn from your host and as you go along. The examples in the chapter will give you an idea what to expect as you embark on your exploration of this vast, resource-rich network.

Cost

Since the Internet itself is not a commercial network, the only costs you incur are associated with your access method. If you are affiliated with an institution that is an Internet site or can get an external account with such an institution, your only costs will likely be those of your telephone calls. If you subscribe to a service that acts as an Internet gateway, you may pay a subscription cost to that service.

For More Information

As a loose affiliation of academic, private, and government networks, the Internet has no central administration. The lack of central authority can be a barrier to learning how to get around on the Internet, but the following organizations can help:

- The NSF Network Service Center (617/873-3400) offers the *Internet Resource Guide,* an excellent reference that explains what is available on the Internet.

- The DDN (Defense Data Network) Network Information Center (415/859-3695) also offers information about the Internet.

Mail

SMTP (simple mail transfer protocol) is the way e-mail gets routed around the Internet. SMTP standardizes the mail you type on your computer's text editor. Usually this form of mail delivery is faster than that of the "store and forward" networks like UUCP or BITNET, since SMTP sends mail directly to the recipient. However, sometimes Internet mail needs to go through gateways, and sometimes you need to know which gateway—for example, BITNET.

Many other networks are starting to copy Internet's mail-addressing scheme, which is called *domain addressing* . The generic format for this type of address is `userid@site.domain`. A domain is a large group of computers lumped together according to the way they talk to each other. The UUCP domain, for example, uses the UUCP protocol to communicate. The BITNET domain is similar. Internet domains include *edu* (educational institutions), *com* (commercial sites), *gov* (government sites), *mil* (military sites), *net* (network centers), and *org* (nonprofits and other organizations that don't belong in the other categories). Domains are often broken down into subdomains.

My Internet address, as an example, is `drittner@uacsc1.albany.edu`. The `@` sign separates the user ID—in this case `drittner`—from the address. The user ID plus the destination address describe for each person on the Internet (or for that matter on any of the networks) his or her own personal mailbox. People's addresses can be looked up in directories similar to telephone directories (see "Locating Other Researchers" later in this chapter).

The second part of my address—uacsc1—stands for the University at Albany Computing Services Center Number 1. This is the site, in this case, of a specific VAX machine. Albany designates the subdomain—University at Albany, New York—which is in the edu domain because it is an educational institution.

Internet addressing isn't always that easy. Many of the networks do not yet follow the domain-addressing scheme. And often you need to use gateway addressing—for example, user@site.BITNET or user@site.UUCP. Fortunately, as Internet becomes more popular, software and gateways are being developed to standardize the methods of addressing, making the task easier every day.

File Transfer

One of the most useful features of the Internet is *ftp*, or file transfer protocol. Many sites, regardless of their locations, allow you to access as a local call certain public areas set aside for exchanging reports, files, software, and other information. A special user ID called "anonymous" along with a password (usually "guest"), allows you to log on to those public areas. Figure 6-1 demonstrates how I downloaded a file in minutes from the DDNNIC (Defense Data Network Network Information Center) in Menlo Park, California, to my host computer in Albany, New York. Boldface type represents what I entered into the computer; material in regular type is what appeared on the screen.

FIGURE 6-1

$ prepare tcp ◄——— *Gets my host ready to connect to the Internet*
TCP prepared
PREPARE done
$ ftp nic.ddn.mil ◄——— *Logs me on to the DDN computer*

```
220 NIC.DDN.MIL FTP Server Process 5Z(47)-6 at Mon
15-Oct-90 18:17-PDT
Remote User Name: anonymous
Remote Password: guest
ftp ls  ←——————— Asks for a list of the contents of the remote
                                directory
00DDN-NEWS-INDEX.TXT.1
00NETINFO-INDEX.TXT.2
00README.TXT.1
00SCC-INDEX..1
12..1
2..1
DDN-MGT-BULLETIN.INDEX.1
DDN-NEWS-INDEX.TXT.1
DECODER.C.1
FILEFILE.TXT.1
FYI..1
FYI-INDEX.TXT.2
IEN-INDEX.TXT.1
KERMIT.DIRECTORY.1
NETINFO..1
NETINFO-INDEX.TXT.1 ←—— I select this file to transfer to my host.
NETPROG-INDEX.TXT.1
NICGUEST.DIRECTORY.1
PROTOCOLS-INDEX.TXT.1
PUB..1
RFC-INDEX.TXT.1
TACNEWS.DIRECTORY.1
ftp> get netinfo-index.txt.1 ←—— Tells my host to download this file
9902 bytes transmitted in 1 seconds (9902 bytes/s)
ftp> quit
                            \
                             The file is now on my host computer in my
                             directory for access.
```

Downloading a file on the Internet using the file transfer protocol (ftp).

Telnet

Telnet, also known as "remote virtual terminal," allows you to log on to a computer on the Internet as though you were directly connected to it, in real time, regardless of its location. You can access a database at another site, or you might want to peruse a library card catalog or use a supercomputer that is connected to the Internet.

Logging on to a site with telnet is as simple as typing `telnet site name`. The example in Figure 6-2 demonstrates gaining access to the WHOIS database maintained by the Defense Data Network (DDN) Network Information Center (NIC). The database lists many of the Internet sites that are registered with NIC. Figure 6-2 illustrates telnet procedures for looking for the Albany and NIC (CA) Internet sites.

FIGURE 6-2

```
$ prepare tcp          ←────── Gets my host ready to connect to the Internet

TCP prepared
PREPARE done

$ telnet nic.ddn.mil ←────── Logs me on to the DDN Computer
Trying...
Connected to nic.ddn.mil.
Escape character is '^]'.
Interrupt remote process character is '^C'.
*-DDN Network Information Center-
*
* For TAC news, type:                    TACNEWS <return>
* For user and host information, type:   WHOIS <return>
* For NIC information, type:             NIC <return>
*
* For user assistance call (800) 235-3155 or (415) 859-3695
* Report system problems to ACTION@NIC.DDN.MIL or call
  (415) 859-5921
```

```
SRI-NIC, TOPS-20 Monitor 7(21245)-4
The system will go down Thu 18-Oct-90 6:00pm until Thu
18-Oct-90 9:00pm for preventive maintenance
@whois
```
◄──────────────── *Asks for user and host information*
```
SRI-NIC WHOIS 3.5(1090)-1 on Mon, 15 Oct 90 18:00:20 PDT,
load 3.53

  Enter a handle, name, mailbox, or other field, optionally
precede by a keyword, like "host sri-nic". Type "?" for
short, 2-page details, "HELP" for full documentation, or
hit RETURN to exit.
---> Do ^E to show search progress, ^G to abort a search or
output
Whois: host uacsc1
```
◄──────────── *Asks for information about the Albany site, New York, Internet*
```
[No name] (UACSC1)

  Hostname: UACSC1.ALBANY.EDU
  Nicknames: RACHEL.ALBANY.EDU
  Address: 128.204.1.2
  System: VAX-8650 running VMS

  Coordinator:
    Leibson, Barry (BL33) LEIBSON@LEAH.ALBANY.EDU
    (518) 442-3711


  Record last updated on 01-Jul-87.


  No registered users.
Whois: nic
```
◄──────────── *Asks for information about the NIC interest site*
```
DDN Network Information Center (NIC)      NIC@NIC.DDN.MIL
  SRI International
  333 Ravenswood Avenue
  Menlo Park, CA 94025
    DDN user assistance    (800) 235-3155   NIC@NIC.DDN.MIL
```

```
                             (415) 859-3695
     Computer Operations     (415) 859-5921     ACTION@NIC.DDN.MIL
     WHOIS updates, user registration
REGISTRAR@NIC.DDN.MIL
     Host changes and updates
HOSTMASTER@NIC.DDN.MIL
     Suggestions
SUGGESTIONS@NIC.DDN.MIL


     Record last updated on 10-Oct-89.


No known members of this organization.
```

Using telnet to log on to a remote database.

Internet Resources

Resources available to you on the Internet include:

- mailing lists,
- supercomputer centers,
- online libraries,
- scientific databases, and
- email access to other environmental researchers and activists.

You have to get proper clearance to access some of those resources, although most areas on the Internet have public-access areas that give you a taste of what is available and information about gaining deeper access.

Mailing Lists

Like FidoNet, Usenet and BITNET, the Internet enables people to discuss various issues with each other by means of "mailing lists." Internet mailing lists are similar to Usenet's newsgroups, FidoNet's echoes, and BITNET's mailing lists. In fact, some of the mailing lists are shared among the networks; if you post a message in one of those lists, it is automatically placed in the other networks' lists.

Appendix B lists Internet mailing lists that pertain to study of the environment and describes how to subscribe to them. Subscribing to the Internet lists is similar to subscribing to BITNET mailing lists; you send a request to a specified address.

Two good sources of information on mailing lists, electronic journals, and newsletters are the *Directory of Electronic Journals and Newsletters* by Michael Strangelove and the *Directories of Academic E-Mail Conferences* by Diane Kovacs. The *Directory of Electronic Journals and Newsletters* enumerates more than 500 scholarly mailing lists, 30 journals, 60 newsletters, and other resources available over the Internet, BITNET, and affiliated networks. The titles include *News of the Earth, South Florida Environmental Reader,* and *Energy Research in Israel Newsletter.* Each entry shows how to subscribe on both BITNET and the Internet. The directory is available free online by sending two commands to the listserver:

```
TELL LISTSERVE AT OUOTTAWA GET EJOURNL1 DIRECTORY

TELL LISTSERVE AT OUOTTAWA GET EJOURNL2 DIRECTORY
```

A printed version of the directory can be obtained for $20 by writing to the Office of Scientific and Academic Publishing, Association of Research Libraries, 1527 New Hampshire Ave. NW, Washington, DC 20036; 202/232-2466.

The *Directories of Academic E-Mail Conferences* contains five files that cover computer-mediated discussions in the social sciences, the humanities, agriculture, biology, environmental and medical sciences; astronomy, chemistry, engineering, geology, mathematics, physics, and general academia (including academic freedom), administration, business, and grants. A sample of titles includes *ECONET*; *ENVST*, the Environmental Studies Discussion list; *AQUIFER@IBACSATA*, a discussion list on pollution and groundwater recharge; *ANIMAL-RIGHTS*; *GREEN*, a discussion of the Green movement; and *RECYCLE*, a discussion on recycling in practice. Each entry shows how to subscribe on both BITNET and the Internet. The directory is available free online by sending a command in the format

```
GET filename filetype to LISTSERV@KENTVM
```

via interactive or e-mail message.

The file name and file types are:

```
ACADLIST FILE1
ACADLIST FILE2
ACADLIST FILE3
ACADLIST FILE4
ACADLIST FILE5
```

You can also get the directory through anonymous ftp from ksuvxa.kent.edu:

Type: `ftp ksuvxa.kent.edu`
signon: `anonymous`
password: `your userid`
type: `GET filename.filetype`
for each file you want, type DIR to see the list on the server. When finished type `quit`

Supercomputers

If you have proper clearance, you may be able to access Internet at a number of computer centers across the country. Many of those facilities offer bulletin boards, workshops, online documentation, 24-hour access, monthly newsletters, system consultants, and even on-site work space. For more information, contact the computer centers directly. They include

- Air Force Supercomputer Center at Kirtland AFB
- Center for Theory and Simulation in Science and Engineering, Cornell National Supercomputer Facility
- John von Neumann National Supercomputer Center
- National Center for Atmospheric Research
- National Center for Supercomputing Applications
- National Energy Research Supercomputer Center
- Northeast Parallel Architecture Center
- Ohio Supercomputer Center
- Pittsburgh Supercomputer Center
- San Diego Supercomputer Center
- U.S. Army Ballistic Research Laboratory
- University of California at Berkeley
- Supercomputing Service, University of Calgary
- Center for Experimental Research in Parallel Algorithms, Software End Systems
- University of Texas System Center for High-Performance Computing
- North Carolina Supercomputing Center
- University of Arizona Supercomputing Center
- UCLA Office of Academic Computing

Online Libraries

More than 50 online library catalogs are available through the Internet, from some of the finest libraries in the country. Using the telnet and ftp protocols, you can access millions of published reference sources for your research. Catalogs from the following libraries are available:

- Auburn University (Alabama)
- California Polytechnic State University, San Luis Obispo, Robert F. Kennedy Library
- University of California Online Catalog
- University of California, Berkeley
- CARL (Colorado Alliance Of Research Libraries)
- University of Delaware Libraries
- LUIS (Library User Information Service, Florida)
- Emory University Libraries Online Public Access Catalog (Georgia)
- University of Hawaii at Manoa Online Catalog
- Northwestern University Library (Illinois)
- University of Chicago
- University of Illinois at Chicago
- University of Illinois, Urbana
- Purdue University (Indiana)
- University of Notre Dame Library (Indiana)
- University of Kansas Library
- University of Maine System Library Catalog
- AIM (Access to Information about Maryland)

- Umcat (Online Catalog for University of Maryland Libraries)
- Boston University
- The University of Michigan
- Michigan State University Libraries
- Wayne State University (Michigan)
- University of Minnesota Library System
- University of Missouri
- Dartmouth College Library Online Catalog (New Hampshire)
- Princeton University Library (New Jersey)
- The University of New Mexico
- New Mexico State University
- Cornell University (New York)
- New York University
- Rensselaer Polytechnic Institute Folsom Library (New York)
- State University of New York At Binghamton
- Case Western Reserve University (Ohio)
- Kent State University (Ohio)
- Ohio State University
- University of Oregon
- Carnegie-Mellon University (Pennsylvania)
- Pennsylvania State University
- Lehigh University (Pennsylvania)

- University of Pennsylvania
- University of Pittsburgh
- Brown University (Rhode Island)
- University of Tennessee
- Vanderbilt University (Tennessee)
- Texas A&M Library System
- University of Texas at Arlington Library System
- University of Texas at Austin Library System
- University of Texas at Dallas
- University of Utah Marriott Library
- Old Dominion University Library (Virginia)
- Virginia Polytechnic Institute and State University
- Virginia Commonwealth University Library Services
- University of Wisconsin Library Catalogs, Madison and Milwaukee Campuses

Procedures for signing on to the libraries can be found in an excellent article entitled "Internet—Accessible Library Catalogs and Databases," by Art St. George. You can get a copy by sending an interactive command or mail message to

```
listserv@bitnic get internet libs
```

The article is also available on the MNS Online BBS; call by modem 518/381-4430.

Online Scientific Databases

You can retrieve research information using email, telnet, or ftp from several specialized scientific database archives on the Internet:

GenBank Server
This server, sponsored by the Institute for Molecular Biology, provides protein information resource (PIR) protein sequence database entries and software for molecular biologists. Other files are available, and you can search by e-mail for a protein sequence. The server is accessed through Internet, BITNET, and UUCP. This server also includes the Matrix of Biological Knowledge database.

COSMIC
Over 1200 programs are distributed here including many for physics and engineering programs. This database is funded by NASA.

IuBio Archive for Molecular and General Biology
Contains software programs and data for biologists, chemists, and other scientists.

PENpages
This service provides thousands of publications on research-based agriculture and consumer issues. It is supported by Penn State's College of Agriculture.

LiMB Database
The Listing of Molecular Biology database contains information on molecular biology databases. Accessible through e-mail only.

Southwest Research Data Display and Analyses System
For people doing research on space physics, magnetospheric physics, and the upper atmosphere. Contains data from the Dynamic Explorer Satellites 1 and 2.

Locating Other Researchers

Several databases on the Internet contain e-mail addresses, postal addresses, phone numbers, and other information about

Internet researchers. Those databases are called the White Pages and can be accessed through e-mail or telnet. Five services can search the White Pages with a single query:

- NASA Ames Research Center (Electronic Phone Book)
- DDN Network Information Center Service (whois)
- NYSERnet/PSI (White Pages Pilot Project)
- CREN/CSNET User Name Server (ns)
- Bucknell University (Knowbot Information Service)

By using e-mail, ftp, and telnet, you can share information with fellow researchers who are on participating networks on the Internet and on gateway services that take mail to other networks such as BITNET, UUCP, and FidoNet. Those networks and services include but are not limited to:

- CICnet (Michigan)
- NYSERNet (New York)
- Sesquinet (Texas)
- USAN (Colorado)
- Westnet (Colorado)
- Los Alamos National Laboratory Integrated Computing Network (New Mexico)
- NASA Science Network (California)
- PREPnet (Pennsylvania)
- SURAnet (Maryland)
- UUNET (Virginia)
- NORDnet (Sweden)
- CSNET (Massachusetts)

- Commercial Mail Relay (California)
- Terrestrial Wideband Network (Massachusetts)
- International Collaboration Network (Massachusetts)
- CONCERT Network (North Carolina)
- SWITCH (Switzerland)
- NevadaNet (Nevada)
- BARRNet (California)
- NorthWestNet (Colorado)
- SUNET (Sweden)
- THEnet (Texas)
- JvNCnet (New Jersey)
- ILAN (Israel)
- ESnet (California)
- WVNET (West Virginia)
- CERFnet (California)
- Los Nettos (California)
- MRNet (Minnesota)
- NASAmail Gateway (California)
- NCSAnet (Illinois)
- NEARNET (Massachusetts)
- NSFNET (Michigan)

A list of all Internet domains and subdomains and their addresses can be found in *A Directory of Electronic Mail* and *The User's Directory of Computer Networks* (see Appendix E).

PART III

Electronic Bulletin Boards

> *The quest for knowledge has to be somewhat opportunistic; we must take it when and where we can.*
>
> Paul B. Sears
> *An Introduction to Ecology*, 1962

CHAPTER 7

Electronic Bulletin Boards

Bulletin board services (BBSs) are operated by individuals, businesses, nonprofit organizations and government agencies for the purpose of sharing information—often at no cost or low cost to the user. Estimates on the number of private bulletin boards in the United States range from a few thousand to more than 100,000.

Bulletin boards evolved during the late seventies as individuals used their own personal computers to conduct public conferencing and debate and to share electronic mail and programs among friends. In the early eighties, bulletin boards quickly became popular with computer user groups as an ideal method of sharing public-domain programs with their memberships. Bulletin boards boomed during the mid eighties as businesses and government agencies found them an attractive method of storing and retrieving large amounts of information.

Bulletin boards generally cater to people with a particular type of computer (for example, IBM PC, Amiga or Macintosh) or a particular interest (for example botany, economics or coin collecting). You can participate in most bulletin boards around the clock, seven days a week at no cost other than the cost of your telephone calls. Some bulletin board operators (SysOps, short for "system operators") request donations from regular users; donors are usually rewarded with additional online time.

Because bulletin boards operate at the whim of their owners and at tremendous personal cost (unless government or business sponsored), they tend to have very short life spans, often as brief as a few months. Boards that have lasted more than a year or two usually have an active membership base, are blessed with a great deal of online expertise, and provide timely information.

Each board takes on the flavor of its sponsor and its membership. One may be popular for its multitude of files for downloading, while another may be popular for the special conferences it carries or because of a special interest of the membership. Although bulletin boards employ different user interfaces, determined by the type of bulletin board software used by the SysOp, all follow similar methods of operation.

This chapter explores:

- connecting to a bulletin board for the first time,

- the basic features and procedures common to every bulletin board, and

- a list of bulletin boards useful for environmental research and discussion.

Connecting to Bulletin Boards

Exploring each bulletin board is a unique adventure, since every board has its own personality and idiosyncracies. This section tells you how to find a bulletin board that interests you and takes you step-by-step through a typical process of accessing a bulletin board for the first time.

It is common for people to panic when they first log on to an unfamiliar BBS. When you first log on to one you may feel as if you entered a deep hole in space. You don't know where the selections will take you. But don't be afraid to explore every section of a board—you won't break anything! Remember you can always turn off your computer without harming anything.

Finding a Bulletin Board

The bulletin boards listed in this chapter are of interest to environmental researchers. If you have never logged on to a bulletin board before, however, you might consider finding one in your own city for your first try. That way, you can take your time and practice before you spend money on long-distance calls.

Each issue of *Vulcan's Computer Monthly* includes a state-by-state list of bulletin boards that are available to the public. To get a copy call 800/874-2937. Many local computing newspapers also publish lists of bulletin boards, and local computer user groups and computer stores can often make recommendations.

Getting Ready to Log On

Once you have found the name and the phone number of a BBS that interests you, you set up your communications software program to communicate with that BBS. Follow the instructions that came with your program to change the communications parameters of the software to match those of the board you wish to call. For example, if the BBS requires parameters of 8 data bits, no parity, and 1 stop bit (usually indicated as "8-0-1" in a BBS listing), set the parameters of your software program to match. Figure 7-1 shows the communications setting parameters of the FreeTerm communications software program, as an example.

Follow the instructions of your software to tell the computer to dial the telephone number of the BBS. You often do this by typing the command `ATDT` followed by the telephone number —for example, `ATDT 3814430`. If you are dialing long-distance, be sure to include the area code, preceded by a 1 if necessary.

FIGURE 7-1

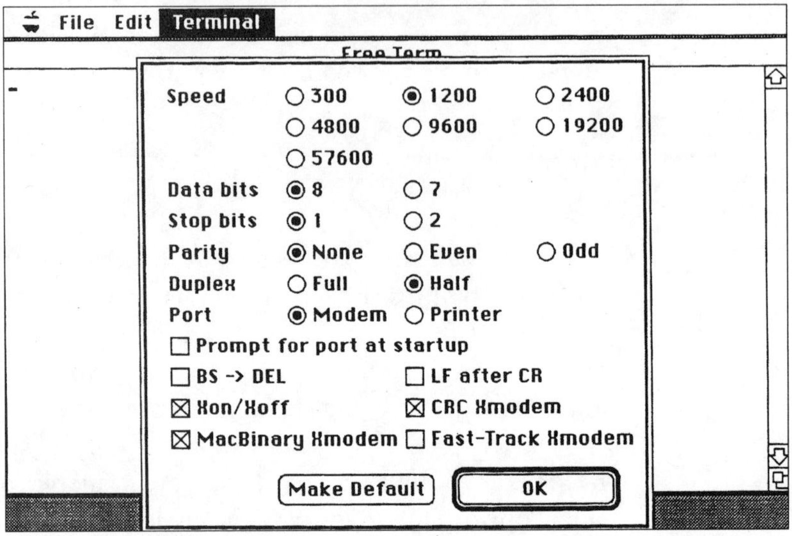

[Screenshot of FreeTerm Terminal settings dialog:
- Speed: ○ 300 ● 1200 ○ 2400 ○ 4800 ○ 9600 ○ 19200 ○ 57600
- Data bits: ● 8 ○ 7
- Stop bits: ● 1 ○ 2
- Parity: ● None ○ Even ○ Odd
- Duplex: ○ Full ● Half
- Port: ● Modem ○ Printer
- ☐ Prompt for port at startup
- ☐ BS -> DEL   ☐ LF after CR
- ☒ Xon/Xoff   ☒ CRC Xmodem
- ☒ MacBinary Xmodem ☐ Fast-Track Xmodem
- Buttons: Make Default, OK]

Setting communications parameters with the FreeTerm communications software program.

Your First Online Session

Connecting

If you have dialed successfully, you see the word CONNECT on your screen. Hit the RETURN or ENTER key once or twice to initiate the BBS. Some bulletin boards prompt you to hit the ESC key to let the BBS know that a human, rather than another computer, is calling in.

What you see next depends on which BBS software the SysOp uses. In the following examples, I demonstrate several types of BBS software to give you a flavor of the various "looks" of BBSs you may encounter in your explorations.

Welcome

A "welcome" screen displays the name of the board; some rudimentary graphics; basic information like baud rate, bits and parity; the name of the SysOp; and other pertinent information the SysOp wants you to know. Figure 7-2 shows a typical welcome screen.

FIGURE 7-2

```
HIT CONTROL-C TO STOP WELCOME MESSAGE
Welcome to ...
*************************************************************
MNS MNS MNS MNS MNS MNS MNS MNS MNS MNS MNS MNS MNS MNS
*************************************************************
     *                    MNS ONLINE                        *
     *                   BULLETIN BOARD                     *
     *               (Formerly The MECC BBS)                *
     *                Founded in April, 1985                *
     *                   MUG NEWS SERVICE                   *
     *                    Located in the                    *
     *           Capital District of New York State         *
     *         Albany, Schenectady, Troy, and Saratoga      *
     *      Serving the Worldwide Apple User Group Community*
     *                  Modem (518) 381-4430                *
     *                      2400 baud                       *
     *                        8-N-1                         *
     *              24 hours a day, 7 days a week           *
     *                   SYSOP: Don Rittner                 *
     *                    FidoNet 1:267/102                 *
     *              We are connected to the World!          *
*************************************************************
Providing news and services to the Apple User Group
Community around the Globe
*************************************************************
MNS ON-LINE runs on a Mac+, 20 meg Dataframe Hard Drive,
and 2400 USR Modem
*************************************************************
Reach Don also on America Online, AFC MNS; GEnie, MNS;
CompuServe, 70057,1325; EcoNet, DRittner; BitNet,
drittner@ALBNYVMS; Internet, drittner@uacsc1.albany.edu;
The Well, drittner
```

```
*************************************************************
MNS MNS MNS MNS MNS MNS MNS MNS MNS MNS MNS MNS MNS MNS
*************************************************************
May your online journey be a pleasant one!
Address email to Don as SYSOP.
```

Welcome screen of a bulletin board that uses Second Sight BBS software.

Registering

Your first time on any BBS, you are prompted for your name and asked to create a password. Some boards give you a password that you can change later. Remember to keep your password secret, but write it down so you don't forget it. Some boards let you use a handle instead of your real name; a handle is simply an alias. Figure 7-3 demonstrates registering with a bulletin board service.

FIGURE 7-3
```
Your FIRST name: don
Your LAST name: rittner
Don Rittner? [Y,n]: y
Wait ...

Welcome to the CAG BBS, the bulletin board of the Computer
Applications Group for the College of Agriculture.
Please enter the CITY and STATE from where you are calling
and then pick a PASSWORD.
Where are you calling from? albany ny
Albany Ny? [Y,n]: y
Pick a password: Test
Test? [Y,n]: y
```

Registering with a BBS that uses the FIDO BBS program. Material entered by the user is boldfaced.

Setting Up Your Monitor Display

As a first-time caller, you are presented with options for how to display words and graphics on your screen. Choices often include

- whether you prefer color graphics,
- whether you want nulls turned on or off (nulls are time delays to skip a line and are useful for printing),
- whether you want ANSI screen commands to provide color,
- what your downloading protocol preferences are,
- whether your terminal can display lowercase characters, and,
- whether you want linefeeds, (linefeeds make sure your display does not type over itself).

You configure the display to your liking and to fit your software and computer. You only need to do this the first time you log on. As long as you remain a member of the BBS, it remembers your default settings. Figure 7-4 shows how the settings are chosen on one BBS.

FIGURE 7-4

```
CAN YOUR TERMINAL DISPLAY LOWER CASE ([Y]/N)? Y
UPPER CASE and lower
* Ctrl-K(^K) / ^X aborts. ^S suspends ^Q resumes *
    Graphics options available are:
N one  means you want only typeable text, like this
       sentence.
A scii uses all 255 IBM characters. Most IBMPC compatibles
       can handle ASCII graphics; yours does if you see
       characters ♥ (heart) and ─── (straight line).
```

C olor uses all 255 IBM characters plus the ANSI screen commands for color & sound. For this, your communications package must support ANSI. Yours does if this => 05;32mblinks in green.00m <=

  Preferences can be changed with the U tilities command, G raphics.
GRAPHICS for text files and menus
Change from N to N)one, A)scii-IBM, C)olor-IBM, H)elp ([ENTER] quits)? **n**
Text GRAPHICS: None
Do you want COLORIZED prompts ([Y],N)? **n**
00;37;40mHighlighting Off

* Ctrl-K(^K) / ^X aborts. ^S suspends ^Q resumes *
File transfer protocol: The method you want to use to transfer files to RBBS ("upload") or to obtain files from RBBS ("download").

TIPS:
1. Select the N)one protocol option if you expect to use a variety of protocols. RBBS will ask for your preference before each file transfer.

2. The protocol you want RBBS to use must also be used by your communications package. Protocols available in RBBS include Ascii (no error checking), Xmodem, xmodem/CRC, Ymodem, and Zmodem. Xmodem is the most widely supported, Ymodem is faster, and Zmodem is best of all.

3. You can override the default by specifying the protocol on the same D)ownload or U)pload command line. e.g. "D;SD.COM;C" downloads file SD.COM in xmodem/CRC, no matter what the default protocol is.
Default Protocol

```
A scii - for text files only
Press Any Key to continue
B atch ymodem (dsz)
C rc xmodem (128 byte blocks)
X modem (128 byte blocks)
Y modem (xmodem with 1024 byte blocks)
Z modem (Batch. The best and fastest.)
N one - cancel

Select Protocol? x
PROTOCOL: X modem (128 byte blocks)

TurboKey: act on 1 character command without waiting for
[ENTER]
Want TurboKeys (Y/[N])? y
TurboKey On
Logging DON RITTNER
RBBS-PC CPC17.2B/0806 NODE 1, OPERATING AT 2400 BAUD,N,8,1
```

Choosing display settings on a BBS running on the RBBS-PC program.

Getting Validated

Many bulletin boards ask you to fill out a validation survey before you are granted full access. This allows the SysOp to know who is using the board and to verify that you are who you say you are. The online form asks for your full name, address, phone number, and other pertinent information.

Boards often deny access if you don't provide the information because, unfortunately, some people upload viruses and conduct other online abuses. So don't be offended if you are asked for some personal information. Expect a phone call from the SysOp.

You may encounter a delay in gaining access to a BBS until your information is verified. Some services let you on the board immediately but deny downloading privileges or participation in certain conference areas until verification is complete. Several boards have an automatic callback feature: the BBS calls your computer after you hang up and asks you to type in your password for verification. Still other bulletin boards have no restrictions; they will let you on with full privileges right after you complete a survey. Most of the boards featured in this book allow you access immediately.

Before you receive final approval, SysOps also require you to agree to a waiver that sets the terms and conditions for use of the board. That procedure is designed to protect the SysOp from lawsuits that might arise if, for example, a user's private mail were made public either by mistake or on purpose. It is important to note here that electronic mail is different from private mail sent through the U.S. Postal Service. The operator of a BBS can read your mail if it is necessary (although SysOps generally do not read your mail without good reason—such as a problem with software or a message being mistakenly sent to a public conference). If you do not agree to the terms and conditions displayed online (usually by typing AGREE), you will probably be denied access. An online survey is shown in Figure 7-5.

FIGURE 7-5

```
If you answer the following questions truthfully, and read
the prompts carefully, you will be validated within 24
hours. If you do not call back within 7 days of your
initial logon your account will be automatically deleted.
In order to maintain your account here you must call at
least once a month if you do not, you will be automatically
deleted.

  If this sounds like a BBS you would like to be a Member
of, answer "Y" or press ENTER at the next prompt to
continue the logon procedure. A response of "N" will log
you off immediately.

Continue to register for membership? Yes
```

```
Enter your name or the handle
you wish to use on this system : DON RITTNER


Now please enter your REAL
first AND last name: don rittner


                    ###-###-####
Enter your VOICE phone number: 222-333-4444


Please enter a password, 3-8 characters: Test


Your sex (M,F) M


In which month you were born (1-12)? 9


On which day of month you were born (1-31)? 8


Of which year? 1948


1. IBM PC (8088)
2. IBM AT (80286)
3. IBM 80386
4. IBM PS/2
5. Apple 2
6. Apple Mac
7. Commodore Amiga
8. Commodore
9. Atari
10. Other
Enter your computer type: 6


How wide is your screen (<CR>=80)? 80


How tall is your screen (<CR>=25)? 25
```

Does your system support ANSI graphics? **No**

```
1. Name          : DON RITTNER
2. Real Name     : don rittner
3. Phone No.     : 222-333-4444
4. Sex           : M
5. Birthdate     : 09/08/48
6. Computer type : Apple Mac
7. Screen size   : 80 X 25
8. Password      : test
Q. No changes.
```

Which (1-8,Q) **Q**

One moment please...

Your Number is 48
Your Password is "test"

    * * * * * * * *

Please write down this information, and re-enter your password for verification.

    * * * * * * * *

Your Password: **XXXX**

 You will now be asked to send a message to the Sysop. You may consider this an application for access. In it you should include the following:

  1) Where did you hear of this BBS?
  2) What would you like to find here?
  3) Your interests, hobbies, and/or any other comments
    you wish to make.

If you do not complete this message you will be automatically disconnected and deleted. After you complete

```
this message, hit "G" and read all the files in the section
entitled "About this BBS......"

E-mailing the Sysop #1


      [-------------------------]
Subject: bbs
Enter message now, max 80 lines.
Enter /S on a new line to save or /HELP on a new line for
help.

Hi This is a request for validation.
don rittner


/s
Saving your message, Don Rittner...
Mail sent to Brian #1
```

The online validation survey of a bulletin board running WWIV BBS software.

Time Limitations

Perhaps the most important information you receive the first time you use a board is how much time you're allotted and how many accesses per day you are allowed. Time allotments average from one half to one hour a day, and access to the board is generally limited to one to three times a day.

SysOps have good reasons for restricting access to their boards. It isn't uncommon for a 15 year old to spend an entire day downloading every game from a board, not contributing anything in return.

Many SysOps reward you with more time or access if you are a frequent participant in public conferences or if you upload files to the board. Some SysOps use a ratio to determine how much time to grant (for example, you receive an extra 30 minutes if you upload five files).

Bulletin Board Features

All bulletin boards offer certain basic features, which are described in this section. However, the features and the procedures for using them differ from board to board, so examples are presented for illustration only. See Appendix D for a log of a complete BBS session.

The Main Menu

The main menu of a bulletin board is the central point of navigation. Figure 7-6 shows a typical main menu. Note that to get help on this bulletin board, you can type the letter H at the prompt.

FIGURE 7-6

```
           *******************************
                       M A I N   M E N U
           *******************************
[E]nter a Message      [A]nswer Questions   [H]elp
[F]iles Subsystem      [K]ill a Message     [B]ulletins
[J]oin Conferences     [G]oodbye            [P]ersonal Mail
[C]omment              [V]iew Conference    [Q]uit to other
[R]ead Messages            Messages             Subsystems
[S]can Messages        [I]nitial Welcome    [X]pert on/off
[U]tilities Subsystem  [O]perator Page      [W]ho else is on
[?]List Functions      [T]opic of Msgs      [U]sers log

MAIN command <?,A,B,C,D,E,F,H,I,J,K,O,P,Q,R,S,T,U,V,W,X>! **P**
Checking messages in MAIN....
NEW Mail for YOU (* = Private)
```

The main menu of a bulletin board running RBBS-PC BBS software. (User input is boldfaced.)

Mail

As soon as you sign on, a BBS looks for any mail addressed to you. Don't expect any mail the first time, although some SysOps will send a help file or a welcome message to first-time callers. A typical BBS mail section looks like the one in Figure 7-7.

FIGURE 7-7

```
              RBBS-PC  MESSAGE  SYSTEM
     ~~~~~~~~~~~~~~~~~~~~~~~~~~~~~~~~~~~
    -- C O M M U N I C A T I O N S --  -- UTILITIES --
    - ELSEWHERE --  PERSONAL MAIL    SYSTEM COMMANDS
    [E]nter a Message    [A]nswer Questions    [H]elp
    [D]oors Subsystem    [K]ill a Message      [B]ulletins
    [J]oin Conferences   [F]iles Subsystem     [P]ersonal Mail
    [C]omment            [V]iew Conferences    [G]oodbye
    [R]ead Messages      [I]nitial Welcome     [X]pert on/off
    [Q]uit to other      [S]can Messages       [O]perator Page
       Subsystems        [?]List Functions     [T]opic of Msgs
    [U]tilities Subsystem


    MAIN command <?,A,B,C,D,E,F,H,I,J,K,O,P,Q,R,S,T,U,V,W,X>! e

    To [A]ll,S)ysop, or name? s
    Subject (Press [ENTER] to quit)? hello
    Make message p[U]blic, p(R)ivate, (P)assword protected,
    (H)elp! r
    Sending personal mail to SYSOP


    Type message 19 lines max (Press [ENTER] to quit)

       [-------------------------------------------------]
    1: Hi!, This is a test of an RBBS-PC BBS mail section.
    2: Don Rittner
```

```
A)bort,C)ont,D)el,E)dit,I)nsert,L)ist,M)argin,
R)ev subj,S)ave

Edit Sub-function A,C,D,E,I,L,M,R,S,?! s
Adding new msg # 336.
Receiver will be notified of new mail
```

A typical BBS mail section. (User input is boldfaced.)

Public Message Boards

All boards have public message sections, or "conference areas." In those areas, the entire membership can discuss topics in an open forum, ask questions, post comments or ideas, and carry on debates.

Some boards have one public message section; others have 20 or more. You can read messages forward or backward chronologically, or individually by number, and on some boards you can follow "threads," or discussions on the same topic among several people. You can send a posted message to a specific person in the conference (for example, if you are answering someone's request for information), or you can type All in the destination box so that a message reaches all the members. Figure 7-8 is an example of a message area that carries several Usenet conferences.

File Libraries

Perhaps the most popular areas of bulletin boards are the file libraries. From them you can download programs, spreadsheets, public-domain files, and text reports. File libraries can be excellent sources of environmental and scientific information. Since many boards are discipline-specific, you can find the latest dis-

FIGURE 7-8

```
MESSAGE AREAS

 1. LOCAL BBS SYSTEM MAIL
 4. Space Activists and Settlement
 5. Science
 6. Science Fiction/Fantasy Literature
 7. Star Trek: The Next Generation
 8. Astronomy
 9. Physics
10. TVRO (Satellite TV)
11. Usenet Sci.Space
12. Usenet Sci.Space.Shuttle
13. Usenet Sci.Astro
14. Usenet Sci.Physics.Fusion
15. Usenet Sci.Physics
16. Usenet Rec.Arts.SF-Lovers
17. Usenet Sci.Nanotech

    Message area [Area #, N)ext, P)rior, ?)list]:
```

The message area of a bulletin board running on OPUS BBS software.

cussion, reports, news, and research. Also, you can often find lists of other bulletin boards to explore.

Different boards offer different combinations of downloading protocols, although all can use ASCII and Xmodem. To save space and uploading and downloading time, files are archived, or compressed, using a variety of archiving programs. You need a program to unarchive the downloaded file, unless it is a straight ASCII text file. All boards offer the software that unarchives the files posted on that board.

BBSs are successful only when all users participate. If you have reports or files that could be of use to the board's membership, please upload them. Some boards even take your downloading privilege away if you do not participate in other areas of the board (for example, the message areas) or upload a certain number of files.

File libraries are often broken down into categories, making it easier to find the information you want. Figure 7-9 is an example of how you find and download a file on a BBS (Example shows a BBS using the Maximus-CBCS BBS software):

FIGURE 7-9

```
File area [Area #, '+'=Next, '-'=Prior, '?'=List]: ?
Text files            Course Materials       Physics
 1 SpaceMet info,     31 AppleWorks files    51 Bibliographies
   bbs lists          32 Astronomy           52 News
 2 Conferences,       33 Biology, Health,    53 Reviews,
   events                Medicine               software lists
 3 Space              34 Chemistry           54 Test Bank
   exploration        35 Computer Science    55 Worksheets
 4 Space Science      36 Course Management   56 Demos, labs,
11 Acid rain,         37 Earth Science,         tips
   water quality         Meteorology         57 Software, IBM
12 Computers          38 Foreign Languages   58 Software, Apple
13 Fidonews           39 General Science     59 Software,
   (Newsletter)       40 Math and               Commodore
14 Energy, radon,        Statistics
   arms               41 Miscellaneous       For Teachers
15 U.S. Dept.of Ed.   42 Reading, writing    64 Harbor Project
16 ERIC files         43 Social sciences     65 MESTEP; jobs
17 Health             44 Space Science Ed    66 MassCUE
   Newsletters                               67 EMAT Files
20 Uploaded Files                            68 Project Shine

Software, general
21 Atari ST           23 Commodore
22 APPLE II           24 IBM
```

```
File area [Area #, '+'=Next, '-'=Prior, '?'=List]: 1
The FILES Section
File area 1 ... SpaceMet Info, BBS Lists
FILE - LINE 4:
A)rea Change     L)ocate a file        F)ile Titles
T)ype (show)     G)oodbye (log off)    D)ownload (receive)
S)tatistics     C)ontents              M)ain Menu
?)help
Select: f


Files: [*)new, =)all, or type a partial filename]:
=-=-=-=- =-=-=-=- =-=-=-=- =-=-=-=- =-=-=-=- =-=-=-=- ===
General Forum User Information Textfiles

Users manual, tips on how to use the board, BBS lists, etc.

All files in this area are text files. To read one on the
screen, you use the T)ype command. For example, to read
a file called FILENAME.EXT, enter
      T FILENAME.EXT <return>
(The notation <return> means press the return or enter key.)
=-=-=-=- =-=-=-=- =-=-=-=- =-=-=-=- =-=-=-=- =-=-=-=- -=-=

CONTRIB.DOC    1664 04-28-88* Appeal for $$ contributions to
               SpaceMet/Physics Forum
MANUAL.DOC    31664 07-14-90* Spacemet Manual 5 '89
PRIMER.DOC    24553 07-14-90* SpaceMet Telecomm Primer 5 '89
OPUSER.DOC    87605 12-03-86* Opus users manual (generic,
               not just Forum)
NEEDHELP.TXT  42003 10-10-86* Help for bbs novices.
               Generic, not just Opus
EDUCATOR.LST  21645 06-26-90* Educators interested in
               Telecomm Projects
FILES.FRM    150790 07-22-90* List of all files on FORUM,
               updated weekly
```

```
FILESFRM.ARC  85940 07-22-90* All files on FORUM - ARchived
                                for IBM's only
QUESTION.FRM  12240 04-03-87* Results of FORUM on-line
                                questionnaire
NASA_CAT.DOC  28204 05-17-90* Video & Audio tapes,
                                filmstrips, slides

Bulletin board lists
MEDBBS89.NOV  29681 11-27-89* Medical and scientific bbs
                                list
MASSNET.LST   10956 05-21-90* Bulletin boards in MA -
                                APRIL, 1989
WILDLIFE.BBS   8005 03-16-90* Forestry, wildlife, ecology
                                bbs
GOVT_DC.BBS    9336 07-18-89* Washington area govt. bbs
EDU.BBS       36764 11-02-87* A national list of
                                educational bbs - rev 6/87
EDUCO.BBS       769 06-12-87* Educational bulletin boards
                                in Colorado
HEALTH.BBS    14093 02-10-88* List of Health Related BBS
APPLE.BBS      9947 08-10-87* BBS phone list Mass plus Apple
A-S902.TXT    18535 01-01-80* Space BBS new March 1990
REGION16.LST  27354 05-21-90* New England Fidonet boards.
                                May 18, 1990
BCS-BBS.LST     608 03-28-90* bulletin boards of the BCS
STUDBBS.DOC    2623 05-14-89* Vermont BBS Project

DISTANT.DOC    3727 03-19-88* Handout for Distant Learning
                                conference
NETWORK.ETC   21613 04-07-88* Harvard Univ. COMMON GROUND
                                BBS Report

TGUIDE.ARC    53410 08-03-87* Arced version of Einstein
                                System - all 5 files
TGINTRO.PRN    6696 06-17-87* Introduction
```

```
TGSRSK1.PRN    30339 07-20-87* Search Skills 1
TGSRSK2.PRN    29348 07-20-87* Search Skills 2
TGAPP.PRN      38130 06-17-87* Classroom Applications
TGGLOSS.PRN     9304 08-03-87* Glossary


RES_DIR.DOC   117812 05-22-90* SpaceMet Resource Ctr
              Directory; see also Minds Database
EDLIST07.ZIP    7658 06-26-90*
MED0690.ZIP    11336 06-26-90* Medical, scientific bbs list
A-S903.TXT     25604 06-02-90* Space bbs June 1990


File area 1 ... SpaceMet Info, BBS Lists

FILE - LINE 4:
A)rea Change    L)ocate a file      F)ile_Titles
T)ype (show)    G)oodbye (log off)  D)ownload (receive)
S)tatistics     C)ontents           M)ain Menu
?)help
Select: d


Select a file transfer protocol:
 1)K-Xmodem
 S)EAlink
 T)elink
 X)modem
 Z)modem
 A) Y-Modem G .... (DSZ)
 B) X-Modem ...... (DSZ)
 C) True Y-Modem . (DSZ) (Not 1K Xmodem)
 D) Z-Modem 32 Bit (DSZ) Batch
 E) Z-Modem 32 Bit (DSZ) MobyTurbo(tm) Batch
 Q)uit

Select: x
```

```
File(s) to download? health.bbs

File: HEALTH.BBS
Size: 14093 bytes (111 Xmodem blocks)
Time: Zmodem: 1:00 Xmodem/Telink: 1:20 SEAlink: 1:02

Mode: Xmodem

Begin receiving HEALTH.BBS now or send several CONTROL-X's
to cancel.

Transfer completed. (CPS=176, 73%)
```

Finding and downloading a file on a BBS using Maximus-CBCS software. (User input is boldfaced.)

Databases

A recent development in BBS technology is the ability to go through a board into another program such as a database. Called a "door," this feature is common on IBM PC-based bulletin boards.

Doors often open to online databases, providing you with another inexpensive way to get specialized information. Figure 7-10 shows the use of a door in a board that specializes in botany. The door leads to an online herbarium collection database, which the researcher searches for all records of Wild Blue Lupine.

FIGURE 7-10

```
1 BUF Local Orchids by Township
2 Not Available
3 BUF Local Ferns by Township
4 BUF Local Ferns Exact Locality
5 Types in the Clinton Herbarium
```

```
 6 1800's Herbarium Specimens in BUF
 7 Field Botanical Types Part 1
 8 Field Botanical Types Part 2
 9 1800's Herbarium Specimens in BUF Part 2
10 NFO Herbarium Specimens
11 ASPT Members—addresses and specialties

Enter # of data base to search,
?# (e.g. ?1) for description of data base,
or Q to quit, [CR] = Q :10
Enter search request
:lupinus
(( LUPINUS
 and :perennis
(( LUPINUS \AND PERENNIS
 and :
(( LUPINUS \AND PERENNIS)
 or ( :
(( LUPINUS \AND PERENNIS))

SEARCHING SEARCH RESULTS
1 entry was found that matches your search request.
Current Output Choices
C) Content change, now: Long form
O) Order of retrieval change, now: Oldest first
D) Display change, now: one by one
S) Show entries
A) Another search
Q) Quit searching
C,O,D,S,A,Q [CR] = Show :s
1 of 1 ( 263 characters)
 Lupinus perennis L.
 Leguminosae
 Canada Ontario
 Reg. Munic. Niagara, City of Niagara Falls
```

```
Flora of Queen Victoria Niagara Falls Park.
   Book No."A" Spec. No. 44 "No. 0 "
Southern Limits of Park.
Roderick Cameron
18........ NFO


N,Q,?,P,R,S,U,D,W,A [CR] = Quit:q
SEARCH RESULTS
1 entry was found that matches your search request.
```

Going through a door in a BBS specializing in botany to search an online herbarium collection database. (User input is boldfaced.)

Bulletin Boards for Environmental Research

The following list of bulletin boards is a small sample of the many useful boards available for environmental and scientific research throughout the United States. The boards featured here are operated by individuals, universities, nonprofit organizations, and government agencies. As you explore the boards, you will find documentation, software programs, ecological models, employment opportunities, grant information, economic statistics, news, databases, and conferences. Most significantly, you will meet hundreds of people online with whom you can discuss issues, request help or information, form alliances, and collaborate on projects.

Most of the boards listed here either offer free access or charge a small fee for complete access. To sign on to any of these boards, set your software parameters for 8 data bits, no parity, and 1 stop bit (8-0-1). If you have problems getting online or connecting, try 7 data bits and even parity.

Agriculture/Plant Science

Agriculture Library Forum

The National Agricultural Library's BBS provides an agricultural date book listing upcoming symposia, conferences, and other events. The Agriculture Library Forum (ALF) maintains an excellent list on agricultural bulletin boards, as well as bibliographies and literature on agriculture, an information center, reference publication, a water quality dateline and information on using ALF and the National Agricultural Library. More than a dozen file libraries, whose contents range from agricultural alternatives to special reference briefs, are available on ALF.

> BBS Number: 800/345-5785, 301/344-5496,
> or 301/344-5497
> Baud Rate: 300, 1200, or 2400 bps
> Availability: 24 hours a day, seven days a week
> SysOp: Karl Schneider
> Help Line: 301/344-2113
> BBS Software: RBBS-PC

Compu-Farm BBS

CFBBS operates out of the Farm Business Management Branch in Olds, Alberta, Canada. An online version of its agricultural software directory lists more than 500 agricultural software products for farm or institution use sorted by category, name, and maker. Features of the BBS include the weekly grain news, crop and livestock market reports, and branch publications on topics such as income tax, estate planning, farm computers, farm accounting and record keeping, and land leasing.

> BBS Number: 403/556-4104
> Baud Rate: 300, 1200, or 2400 bps
> Availability: 24 hours a day, seven days a week
> Help Line: 403/556-4243

> SysOp: Bruce Waldie
> BBS Software: RBBS-PC

Rutgers Cooperative Extension BBS
RCEBBS is operated by the Rutgers University Cooperative Extension. Bulletins are available on topics including pathology, floriculture, farmer's markets, pesticides, horticulture, financial news, newsletters, and food technology.

A door area contains the Alternative Farming Systems Literature database, which lists more than 6000 citations, dating from the sixties to the mid eighties, covering all aspects of organic growing and sustainable agriculture.

> BBS Number: 800/722-0335 (New Jersey only) or
>   201/383-8041
> Baud Rate: 2400 bps
> Availability: 24 hours a day, seven days a week
> SysOp: Bruce Barbour
> BBS Software: PCBoard
> Comments: May now be open by subscription only

Integrated Pest Management BBS
IPB BBS operates from the University of Wisconsin Extension in Madison. The board's focus is on sustainable agriculture and integrated pest management.

> BBS Number: 608/262-3656
> Baud Rate: 300, 1200, 2400, 4800, or 9600 bps
> Availability: 24 hours a day, seven days a week
> SysOps: Roger Schmidt and George Rice
> Help Line: 608/262-0170
> BBS Software: RBBS-PC

Air Quality

Support Center for Regulatory Air Models (SCRAM) BBS
This is EPA's air dispersal modeling bulletin board. It is where air quality models are updated and discussed. You can download Fortran source code and compiled versions of all the current regulatory air models from this BBS. It also has conferences, mail, and a downloadable user manual.

> BBS Number: 919/541-5472 (for baud rate 1200 or 2400 bps)
> 919/541-1447 (for baud rate 9600 bps)
> Availability: 24 hours a day except Mondays, 8–12 EST
> SysOp: Hersch Rorex
> Help Line: 919/541-5384

Astronomy/Space

NASA SpaceLink
Operated by the Marshall Space Flight Center and NASA educational affairs, this is an extensive space and science BBS designed for educators. Areas include current NASA news, aeronautics, space exploration, space program spin-offs, classroom material, and NASA educational services.

Teacher and student files cover science fairs, lunar sample educational projects, teacher workshops, and more. Educators can order NASA publications. Classroom materials include software, careers in aerospace, space science activities, NASA facts, posters, and technical and scientific publications.

> BBS Number: 205/895-0028
> Baud Rate: 300, 1200, or 2400 bps
> Availability: 24 hours a day, seven days a week

> SysOp: Bill Anderson
> Help Line: 205/544-0994
> Comments: The first time you sign on, enter `Newuser` as your user name and password.

National Space Society BBS
Sponsored by the National Space Society (NSS) as a service to the public, this board covers space science and technology issues. NSS carries a few Usenet conferences: sci.space, sci.space.shuttle, sci.astro, sci.physics, sci.physics.fusion, sci.naotech. The board also contains special areas for NSS members, including chapter and membership lists. File libraries and bulletins include press releases and articles from NASA, their own NSS Space Hotline news, and more than 20 downloading libraries.

> BBS Number: 412/366-5208
> Baud Rates: 1200, 2400, or 9600 bps
> Availability: 24 hours a day, 7 days a week
> SysOp: Bev Freed
> BBS Software: Opus

Biological Sciences

The BioTron
The BioTron, sponsored by the American Foundation for Biological Sciences (AFBS), announces employment opportunities in the biological sciences before they appear in AFBS's well-respected publication *BioScience*. You can register online for annual scientific meetings and correspond with all AFBS departments. Current issues of *The Forum*, a public policy newsletter, can be read online, and you can download the previous year's issues.

BBS Number: 202/628-2427
Baud Rate: 300, 1200, or 2400 bps
Availability: 24 hours a day, 7 days a week
Help Line: 202/628-1500
BBS Software: TBBS

The National Biological Impact Assessment Program BBS
NBIAP operates from the Virginia Polytechnic Institute and State University. Designed to "facilitate the safe evaluation of the performance of genetically modified organisms in the environment," the board focuses on research and monitoring techniques in biotechnology. You can search more than a dozen databases that cover regulations and guidelines for specific organisms. News sections include current issues in biotechnology. Also included are lists of biotechnology companies, job opportunities, and an international plant biotechnologist directory.

BBS Number: 800/624-2723 (You are allowed ½ hour toll-free every 24 hours) or 703/231-3858
Baud Rate: 300, 1200, or 2400 bps
Availability: 24 hours a day, seven days a week
BBS Software: PCBoard

National Science Foundation BBS
Operated by the National Science Foundation (NSF), the NSF BBS offers bulletins and information on studies funded by NSF in science and engineering. Interesting bulletins on biotechnology, women and minorities in science and engineering, R&D resources in industry, and more can be found on this BBS.

BBS Number: 202/634-1764
Baud Rate: 300, 1200, or 2400 bps
Availability: 24 hours a day
Help Line: 202/634-4250
BBS Software: RBBS-PC

Taxonomic Reference File
The Taxonomic Reference File (TRF) is operated by BIOSIS, a nonprofit organization that produces the *Biological Abstracts* and the *Zoological Records*. TRF offers searchable molecular biology databases that focus on bacteriology, bacterial nomenclature, and molecular biology and culture collections.

> BBS Number: 215/972-6759
> Baud Rate: 300, 1200, or 2400 bps
> Availability: Noon to 8 a.m. Mon. through Fri.;
> 24 hours Saturday and Sunday
> SysOps: Bob Howey, Carol Lock, Keith Pittman, and Yolanda Bryant
> Help Line: 215/587-4917
> BBS Software: RBBS-PC

Botany

TAXACOM
A service of the Buffalo Museum of Science, (TAXACOM) is for botanists and anyone else working in the natural sciences.

Anyone can contribute to TAXACOM's online electronic journal, *Flora Online*. *Clintonia*, the bimonthly magazine of the Niagara Frontier Botanical Society, is also available online, as is the informal newsletter *TAXACOM Cyclopedia*. Also distributed online are *The Bean Bag*, an electronic newsletter for legume specialists, and Biosphere, a BITNET magazine on the environment.

Members can hold minisymposia in conferences that include botanical Latin, bryology and lichenology, offers of positions, online communications, environment, phylogenetics, ornithology, mycology, and many others. A unique service of TAXACOM is a Latin translation service for taxonomists.

Several databases can be searched using Boolean operators:

- the Field Museum of Natural History Botanical Type Photograph database, which has more than 62,000 records;
- ferns and orchids of western New York State;
- historical records in the Clinton Herbarium (1800s); and
- the American Society of Plant Taxonomists/Research Specializations, which has 1176 records.

> BBS Number: 716/896-7581
> Baud Rate: 300, 1200, or 2400 bps
> Availability: 24 hours a day, seven days a week
> SysOp: Richard Zander
> Help Line: 716/896-5200
> BBS Software: FYI-MCD

Climate

National Oceanographic and Atmospheric Administration (NOAA)
This board is provided by the Space Environment Laboratory (SEL) of the NOAA, U.S. Department of Commerce, and includes daily text messages produced by the SEL's Space Environment Service Center (SESC). The Center monitors solar activity in collaboration with the U.S. Air Force and provides advisories and forecasts of that activity and the effects on the near space environment of the Earth. Topics covered include geomagnetic, ionospheric, atmospheric, and space environment effects.

> BBS Number: 301/770-0069
> Baud Rate: 300, 1200, or 2400 bps

Conservation

EnviroNet

EnviroNet, sponsored by Greenpeace, is for anyone interested in peace and environmental issues. This active board contains press releases from Greenpeace, bulletins, new user information, and a list of Greenpeace's North American national and regional offices.

Seven conferences are carried: gossip, disarmament, energy, forests, general, stepping lightly, and toxics. Each is moderated by a different individual. Online discussion with members is available in a special chat area. EnviroNet has an extensive list of files for downloading, covering all aspects of the environment, from grassroots conventions to dioxins. Qualifying groups can access EnviroNet by means of a special toll-free number.

> BBS Number: 415/861-6503
> Baud Rate: 300, 1200, or 2400 bps
> Availability: 24 hours a day, 7 days a week
> SysOp: Dick Dillman

Ecological Modeling

National Ecology Research Center

This board is for researchers modeling fish and wildlife habitats such as the Habitat Evaluation Procedures (HEP) and the In-Stream Flow Incremental Methodology (IFIM). The board is operated by the U.S. Fish and Wildlife Service National Ecology Research Center.

> BBS Number: 303/226-9365
> Baud Rate: 300, 1200, or 2400 bps
> Availability: 24 hours a day, 7 days a week

SysOp: Gene Whitaker
Help Line: 202/343-3245
BBS Software: RBBS-PC
Comments: Call 303/226-9335 for a hardcopy of a user guide.

Environment

Eco System BBS

The Eco System Bulletin Board, based in Pittsburgh, Pennsylvania, is dedicated to issues of environment and economics. This easy-to-use system has conferences, text files, and public domain MS-DOS programs. Eco System contains national job listings for activists, conferences and files on recycling and composting, and a list of environmental organizations and their representatives. All files are downloadable on the first call. Eco System BBS has only one telephone access line, so keep trying.

BBS Number: 412/244-0675
Baud Rate: 1200/2400
SysOp: Mike Shafer

MNS Online

MNS Online is the "official EcoLinking BBS," run by the author of this book. MNS Online carries many environmental and science echoes (for example, Biosphere, Bionews, and EcoNet). It has a file section from which you can download environmental news, reports, and programs. An EcoLinking area provides updates and other information on using online databases for environmental research.

BBS Number: 518/381-4430
Baud Rate: 300, 2400, or 9600 bps
Availability: 24 hours a day, 7 days a week
SysOp: Don Rittner

BBS Software: First Class

Geology

United States Geological Survey BBS
This BBS contains files related to geology, mining, minerals, seismology, and mapping.

> BBS Number: 703/648-4168
> Baud Rate: 1200 bps
> Availability: 24 hours a day, 7 days a week
> SysOp: Joe Kempter
> BBS Software: PCBoard

Hazardous Waste/Toxics/Pesticides

Hazardous Materials Information Exchange
The Hazardous Materials Information Exchange (HMIX) is sponsored by the Federal Emergency Management Agency (FEMA) Technological Hazards Division in Washington, D.C. The board provides a centralized database in which federal, state, local, and private sector users can share information pertaining to hazardous materials emergency management, training, resources, technical assistance, and regulations. More than 30 bulletins on this board cover a wide range of issues related to hazardous materials, community right-to-know laws, EPA grants, chemicals subject to reporting under the Emergency Planning and Community Right to Know Act (available on disk), and hazardous waste education. Bibliographies are also available.

Downloadable files include lists of hazardous-materials-related bulletin boards, toll-free numbers, federal and state agency online databases, (for example, from the Department of Transportation, OSHA, and the EPA), federal and state toll-free technical assistance numbers, and commercial online databases.

A section on organizational contacts lists federal agencies, professional coalitions, trade associations, research centers, environmental groups, and state and local public-interest groups. Also included is an extensive list of educational resources, newsletters, and journals on waste management and federal agency publications.

> BBS Number: 708/972-3275
> Baud Rate: 300, 1200, or 2400 bps
> Availability: 24 hours a day, 7 days a week
> SysOp: Lessie Graves
> Help Line: 800/752-6367, 800/367-9592 in Illinois
> BBS Software: PCBoard

FireNet Leader

The focus of this BBS is on fire fighting and emergency medical treatment. A file library contains reports on hazardous wastes, how to handle toxics in case of a fire, federal and state laws regulating the handling of toxics, environmental groups and coalitions, clean air amendments, EPA rules and regulations, transportation regulations, and toxicology profiles of hazardous materials.

> BBS Number: 719/591-7415
> Baud Rate: 300, 1200, 2400, or 9600 bps
> Availability: 24 hours a day, 7 days a week
> BBS Software: Maximus-CBCS

OSWER BBS

The Office of Solid Waste and Emergency Response (OSWER) is part of the EPA's Technology Innovation Office. The OSWER BBS is a restricted bulletin board that is fully available only to employees of federal, state, and local government agencies and EPS contractors. Others are afforded full access only to bulletins and to the following Special Interest Groups (SIGs) 5 and 7. The full list of SIGs is as follows:

1. Environmental Services Division (for questions and answers on analytical chemistry field analysis and laboratory audits).

2. Ground Water Workstation, for users of the 14 hazardous waste groundwater workstations.

3. Superfund Analytical Services

4. Superfund Removal Program

5. Air/Superfund Coordination, for regional air division technical staff who advise on Clean Air Act requirements at Superfund sites.

6. Incineration and Thermal Treatment.

7. Superfund Innovative Techology Evaluation (SITE)

8. UST Health and Safety Interactive Video Training.

File libraries contain EPA database and subject directories that run on IBM PCs—for example, an annotated bibliography of groundwater publications and databases providing treatability information, a coordinated list of chemicals, a technical support project, and more. OSWER provides more than 40 online bulletins on how to use the board, lists of various laboratories, planning guides for lead battery sites and wood preserver sites, hazardous waste *Federal Register* notices, a hazardous waste library collection database updating program, and a list of EPA regional libraries.

> BBS Number: 301/589-8366
> Baud Rates: 1200, or 2400 bps
> Availability: 24 hours a day, 7 days a week
> Help Line: 301/589-8368
> BBS Software: PCBoard
> SysOp: Dan Powell

RACHEL (Remote Access Chemical Hazards Electronic Library)

This full-text searchable database, sponsored by the Environmental Research Foundation (ERF) of Princeton, New Jersey, contains a storehouse of information about a wide range of problems associated with hazardous materials. The staff will do free computer searches. RACHEL also has the following: (1) national civil dockets of corporate EPA violations, the Bad Actor's Tracker network (BAT), listing corporate waste handlers from 1972–1990; (2) *New York Times* abstracts since 1985; (3) a free hardcopy newsletter, "Hazardous Waste News." The database has full Boolean search capability in nine databases:

- Abstracts from newspapers, magazines (more than 6000 documents)
- New Jersey Department of Health Fact Sheets on Chemicals (724 documents)
- U.S. Coast Guard fact sheets on chemicals (over 1000 documents)
- Information about landfills (15 documents)
- Information about incinerators (16 documents)
- New technologies, recycling, waste avoidance (13 documents)
- Information on particular waste-handling companies (165 documents)
- Sensible public policies (5 documents)
- Reference information about the environment (188 documents)

BBS Number: 609/683-0045
Baud Rate: 1200, or 2400 bps
Availability: 24 hours a day, 7 days a week

SysOp: Peter Montague
Help Line: 202/328-1119. Open 8 a.m. to 7 p.m. EST, Monday–Friday
BBS Software: BRS/Search UNIX
Comments: Call the ERF HelpLine for a four-page user guide and a user's account. You need to sign a form for a free account.

Herpetology

HerpNet

The Satronics TBBS, operational since 1984 added HerpNet to allow those with an interest in reptiles or amphibians to communicate with each other. The national—and often international—callers are usually interested in discussing captive breeding projects, conservation, behavior, cage design, and numerous aspects of the discipline. Professionals, and veterinarians, as well as hobbyists and biology students, are members.

The file area of HerpNet contains a database of herpetology clubs, societies, and organizations across the United States (in text and dBase III format), as well as ASCII text files about care, hibernation, breeding, and the medical management of snakebites and conservation; bibliographies are also available. HerpNet has a searchable database that lets you use keyword to find herpetological materials submitted by other members of the board.

BBS Number: 215/464-3562
for 9600 baud access, call 215/698-1905
Baud Rate: 300, 1200, 2400, or 9600
Availability: 24 hours a day, 7 days a week
SysOp: Mark Miller
Help Line: 215/464-3561
BBS Software: TBBS

Hydrology/Water Resources

Water and Wastewater Network
Cosponsored by the Michigan chapter of the American Water Works Association and the Michigan Water Pollution Control Association, this BBS covers water and wastewater issues. It includes the *Federal Register* abstracts online.

Five bulletin areas offer discussions on water, wastewater, laboratories, and water research. Downloadable files include *Federal Register* abstracts, research files, how to choose a professional consultant, EPA safe drinking water act communications, and EPA publications regarding water issues.

> BBS Number: 517/686-4055
> Baud Rate: 300, 1200, or 2400 bps
> Availability: 24 hours a day, 7 days a week
> SysOps: John DeKam and Randall Fisher
> BBS Software: PCBoard

Medicine/Public Health

St. Joseph's Hospital BBS
This board offers FDA news and releases, the latest AIDS information, employment opportunities, medical news, and Centers for Disease Control reports. Many text files on environmental and public-health-related subjects are available to download, including articles on Chernobyl, diaper safety, surgeon general reports, toll-free numbers for health information, poison control information, the effects of air pollution on lung functions, and more.

> BBS Number: 602/235-9653
> Baud Rate: 300, 1200, 2400, or 9600 bps
> Availability: 24 hours a day, 7 days a week

SysOp: David Dodell
BBS Software: Opus-CBCS

Black Bag BBS
The Black Bag carries science and public health FidoNet echoes such as Science National, National Physics, and Radiology. It is a member of the FidoNet, so you can send e-mail to the Internet and BITNET.

File categories include epidemiology, science programs, a health information newsletter, and astronomy. *HICN News*, the Health Information Communication Newsletter, published weekly by David Dodell at the St. Joseph Medical Center in Phoenix, Arizona, is distributed here, since this BBS is the financial sponsor of the newsletter. The *American Physical Society Bulletin*, a weekly update of political news and recent scientific discoveries, is also available.

The Black Bag BBS is the home of the Black Bag Medical BBS list, which is maintained by the SysOp. It lists over 240 science-related bulletin boards worldwide and is updated quarterly.

BBS Number: 302/731-1998
Baud Rate: 300, 1200, or 2400 bps
Availability: 24 hours a day, 7 days a week
SysOp: Ed Del Grosso
BBS Software: Opus-CBCS

Nature

Skyland
Skyland is designed for nature photographers, nature writers, and nature lovers. Conferences include discussions on the Smokies and related events in western North Carolina, writings, photography, the earth, New Age, and more. File sections include the Smokies, flora and fauna, geography, photography, environ-

ment, and the outdoors. Skyland provides useful lists of other wildlife and natural science bulletin boards.

> BBS Number: 704/254-7800
> Baud Rate: 300, 1200, or 2400 bps
> Availability: 24 hours a day, 7 days a week
> SysOp: Michael Havelin
> BBS Software: PCBoard

Ornithology

Osprey's Nest

The Osprey's Nest, part of the National Birding Line Cooperative, is a bulletin board for birders, naturalists, and conservationists. The board provides a forum for discussion and information exchange on bird-watching and bird feeding in the Washington D.C. metropolitan area. The board serves the local birding community with news of sightings, information on local hot spots, transcripts of local hot lines, and reviews of equipment and books. Although the board is a vast informational resource, the exchange that takes place is social in nature as well; the SysOps have sponsored get-togethers of their members.

> BBS Number: 301/989-9036
> Baud Rate: 300, 1200, 2400 bps
> Availability: 24 hours a day, 7 days a week
> SysOps: Fran and Norm Saunders
> BBS Software: ROBBS

Physics

Spacemet Central/Physics Forum

This board, operated by the Department of Physics and Astronomy at the University of Massachusetts, is primarily for teachers in that state.

Message areas and files cover the topics of space, physics, and astronomy. Educational files cover acid rain, water quality, biology, health, general science, earth science, course management, and chemistry. Elementary and junior high school students can participate in a statewide pen pals program. The board carries several FidoNet echoes on science and the environment.

> BBS Number: 413/545-1959 or 413/545-4453;
> 617/265-8972 in Boston
> Baud Rates: 300, 1200, or 2400 bps
> Availability: 24 hours a day, 7 days a week
> SysOp: Helen Sternheim
> Help Line: 413/545-3697, 413/545-2548
> BBS: Maximus-CBCS

Pollution Abatement

Center for Exposure Assessment Modeling BBS

Operated by the EPA Environmental Research Lab in Athens, Georgia, this BBS is designed to distribute environmental exposure assessment models supported by the Center for Exposure Assessment Modeling (CEAM) and to provide interactive user support. All CEAM models available for personal computers can be downloaded from the CEAM BBS file area, as can the coordinated list of chemicals database (CLC) database. The CLC database lists chemicals slated to be studied experimentally over the next two to three years by eleven EPA research laboratories. Each chemical studied is referenced in the database by the name of the laboratory that will study it, the principal researcher, and the processes to be studied. The database also includes the names of chemicals contained on ten federal regulatory lists. Each chemical in the database is cross-referenced for its presence on the regulatory lists.

Also available on the CEAM BBS are environmental models for aquatic, atmospheric, terrestrial, and multimedia pathways for organic chemicals and metal; all run on IBMs and IBM compatibles. The BBS can be used to post questions concerning the theory, application, and installation of the models; it also offers workshops and provides updates on errors and other information regarding the applications of the models in real life.

> BBS Number: 404/546-3402
> Baud Rate: 1200, 2400, or 9600 bps
> Availability: 24 hours a day, 7 days a week
> SysOp: Shawn Turk
> Help Line: 404/546 3549

Pollution Prevention Information Clearinghouse (PPIC) and Pollution Prevention Information Exchange System (PIES)
PIES is sponsored by the EPA's Office of Environmental Engineering and Technology Demonstration (OEETD) and Office of Pollution Prevention in Washington, D.C. The BBS provides a preliminary source of information on pollution prevention programs and options that may save money, reduce liability, and reduce degradation of the environment.

Bulletins include articles on pollution prevention, waste reduction, and environmental protection. You can find the *Pollution Prevention Newsletter*, employment opportunities, and *Federal Register* notices regarding pollution. Federal and state pollution prevention program descriptions are available in ASCII text format and may be downloaded. The BBS contains many excellent articles on pollution and hazardous waste issues.

Searchable databases on PIES include a calendar of events; federal, state, and corporate program summaries; case study abstracts; general publication abstracts; the PPIC contact list; and legislation summaries.

The PIES mini-exchange contains the UNEP International Cleaner Production Information Clearinghouse (ICPIC), the Ecological Products Workgroup, the Waste Exchange Forum, the

American Institute for Pollution Prevention (AIPP), EPA Region One Association of States and Interstate Agencies, the EPA Region Ten Northwest Regional Roundtable, Federal Pollution Prevention Grants, the Local Government Exchange, the Research Exchange, the Indiana Waste Exchange and the Pollution Prevention Program.

> BBS Number: 703/506-1025
> Baud Rate: 1200, or 2400 bps
> Availability: 24 hours a day, 7 days a week
> Help Line: Myles Morse, EPA Technical Representative, 202/475-7161; EIES Technical Support, 703/821-4800
> BBS Software: PCBoard

Nonpoint Source Information Exchange BBS
The Nonpoint Source (NPS) Information Exchange BBS provides state and local agencies, private organizations, businesses, and concerned individuals with timely NPS information, a forum for open discussion, and the ability to exchange computer text and program files. The BBS is operated by the EPA's Office of Water. More than 100 bulletins and files related to nonpoint source and water pollution issues are available for downloading. A door to an excellent searchable database called Clean Lakes Database allows you to find publications on water issues by word in a title, subject, author, region, or state. The database provides you with a citation and an abstract of the article.

> BBS number: 301/589-0205
> Baud Rate: 1200, or 2400 bps
> Availability: 24 hours a day, 7 days a week
> SysOp: Judy Trimarchi
> Help Line: 301/589-5318
> BBS Software: PCBoard
> Comments: You can call the help line to request a complete user manual and telecommunications fact sheet.

Science Education

National Geographic Kids Network

National Geographic has developed Kids Network for teachers and schools that want to teach students how to use online communications and understand science and environmental issues.

Students research and conduct experiments on an issue, then share their findings online with teammates around the world. A professional scientist on the network works with the students. Each session is a six-week course; currently three sessions available: Hello!, Acid Rain, and Weather in Action. Participants must use Apple II computers. Teachers who enlist in the program receive the Kids Network program disk, a teacher guide, lesson plans, a software manual, activity sheets, student handbooks, wall maps, and other scientific materials. The sessions cost around $400 each. Contact National Geographic at 800/368-2728 for more information.

The Amateur Scientist BBS

A board for people who are interested in science, especially students and other amateurs but including professionals who like to discuss their own and other fields with nonspecialists. Includes The *Journal of Student Research* (*JSR*), devoted to research papers by high school students, tutorial articles, and essays of interest to young scientists. Anyone is invited to submit papers; the board is a place for prepublication review as well as a place to read papers already published in *JSR*. The board maintains a library of programs useful in amateur scientific work and lists of other science-related boards. Also available is the Trans-Pacific Nursery catalog, a file containing descriptions of rare, unusual, and in some cases threatened or endangered plant species.

> BBS Number: 503/843 4214
> Baud Rate: 300, 1200, or 2400 bps

> Availability: 24 Hours a day, 7 days a week
> SysOp: Gerry Roe
> BBS Software: Wildcat

Science Line BBS
Science Line is sponsored by the National Science Teachers Association (NSTA). File libraries contain a wealth of information and files for science teachers in astronomy, communications, earth science, medicine, psychology, general science, science for the handicapped, and more.

The *Science Line Navigation Guide*, a publication for new and inexperienced users of bulletin board systems, walks you through most of the features of Science Line, all the way from getting online to file transfers and automated sessions. The *Navigation Guide* is available for $3.00 ($2.65 to NSTA members) from NSTA Publications, 1742 Connecticut Ave. NW, Washington, DC 20009.

> BBS Number: 202/328-5853, 202/265-4496
> Baud Rate: 2400 bps
> Availability: 24 hours a day, 7 days a week
> SysOp: Alex Mondale
> Help Line: 202/328-5800 ext. 57
> BBS software: RBBS-PC

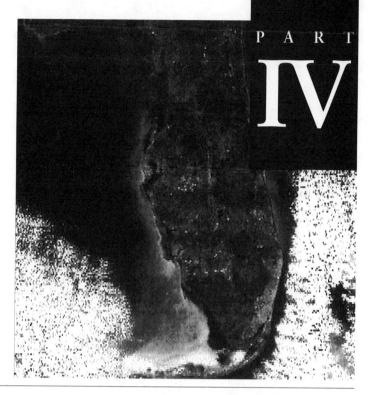

PART IV

Commercial Online Services

Perhaps the most significant recent development in the computer industry has been the growth of personal ownership of computers. The 1989 U.S. Census found that 15 percent of U.S. households own a personal computer—up from 8.2 percent in 1984.

Increased access to personal computers has brought with it familiarity, and as individuals have become familiar with their computers, they have found new uses for them. In 1989, 3.1 million U.S. households that owned computers also had modems.

In response to this expanding interest in telecommunications, several companies developed online services to accommodate personal computer owners. The online information service industry was born.

In 1979, CompuServe became the first commercial online provider to the personal computer user, and the market has continued to grow with the latest entry, America Online, in 1989.

All the providers operate from one or more host computers, have multiuser capability, and charge a fee for access. Providers such as GEnie, The WELL, and CompuServe cater to people with all types of computers, while America Online caters to owners of IBM PC, Macintosh, and Apple II computers. The nonprofit EcoNet caters specifically to the environmental community.

Hundreds of services are available from these online providers. You can shop for clothes, electronics, even food. You can follow stock quotes and business leads. You can read the latest environmental news from wire services such as UPI and AP. You can book reservations on airlines, send private mail to colleagues, participate in public conferences in real time with other members, and join computer SIGs (special interest groups) for your particular computer type or interest. The services also provide file libraries from which you can download and to which you can upload computer files, spreadsheets, and programs. The five information services covered in this section of the book—America Online, CompuServe, EcoNet, GEnie, and The WELL—all have environmental or science-related resources.

Unlike private bulletin boards, these online providers charge an hourly fee for access. Some require a one-time signup fee or a monthly minimum; all have an hourly charge, ranging from as low as $2 an hour to a high of $12.50 an hour.

You may want to explore all the services offered by these online companies, but this book focuses only on areas that pertain to environmental networking.

> *In the long run, it is the sum total of the actions of millions of individuals that constitute effective group action.... Get involved in political action. Otherwise, we shall all eventually find ourselves stranded in space on a dead Spaceship Earth with no place to go and no way to get there.*
>
> Paul Ehrlich

CHAPTER 8

America Online

America Online is a commercial online service for users of IBM PC-compatible, Macintosh, and Apple II computers. Rather than requiring a "generic" communications software package (as do many of the online resources described in this book), America Online is accessed through a graphic software program, customized to your type of computer. Users of all experience levels—from expert to novice—can navigate the service easily using windows, pull-down menus, and the "point and click" convenience of a mouse. Unlike with many online databases and bulletin boards, America Online members do not need to memorize complex commands—or even understand the intricacies of telecommunications—to take advantage of the benefits an online service has to offer.

America Online offers members private electronic mail, message boards covering a variety of special interests, news feeds and business information, shopping online, travel arrangements, "conference rooms" where members can discuss issues in "real time," online classes and nightly "homework help" tutoring sessions, computing support forums, and special interest groups that explore a wide variety of topics, including the environment and sciences.

How to Get Online

America Online membership is open to anyone with an IBM PC-compatible, Macintosh or Apple II personal computer and a modem. To use America Online, you must obtain an America Online software kit, which you can get at no charge by calling 800/827-6364. The software kit includes easy, step-by-step instructions for getting online.

Cost

Online fees are $5 per hour for nonpeak time (evenings and weekends) and $10 per hour for peak time (business hours), for either 1200 or 2400 baud. A monthly membership fee of $5.95 includes one nonpeak hour free per month. Local telephone access is provided through the SprintNet and Tymnet networks.

Organization

America Online services are divided into the following eight subject areas, called "departments" (see Figure 8-1):

- News & Finance
- Entertainment
- Travel & Shopping
- People Connection
- Computing & Software
- Lifestyles & Interests
- Learning & Reference
- What's New & Online Support

FIGURE 8-1

America Online's main menu shows categories of information available. To go to an area that interests you, click the mouse on the appropriate graphic icon.

Getting Around

You can "navigate" through America Online in two ways: you can use the mouse to point and click through a hierarchical "path" of services symbolized by graphics (*icons*), and lists (*menus*); or you can use keyword commands that take you directly to the information you seek. To use a keyword command, you hit the COMMAND and letter *K* keys at the same time. A form pops up prompting you to type in the appropriate keyword. Type it in and press RETURN, and you automatically go to the area you have selected.

Environmental Resources

People in the America Online community use the service for a wide range of reasons: education; computing support; the social interaction of clubs, special interest groups, and professional groups; information; research; and fun. This chapter explores America Online resources of particular interest to EcoLinkers:

- The Environmental Forum
- The online encyclopedia
- News wire services

The Environmental Forum

> Keyword: Earth
> Path: Lifestyles & Interests > Environmental Club

Description

The Environmental Forum is a special interest area found in the Lifestyles & Interests department (see Figure 8-2). Like every America Online club, the Environmental Forum has a host who is responsible for maintaining the club's message boards, conferences, and software library; in this case, the forum host is the author of this book, who goes by the screen names of Host Earth, AFL DonR, and DRittner.

The Environmental Forum is designed for public discussion and debate on any environmental issue. Any member of America Online can become a member of the forum, at no additional charge. It is open to all in the hopes of fostering dialogue among members interested in environmental and scientific issues.

FIGURE 8-2

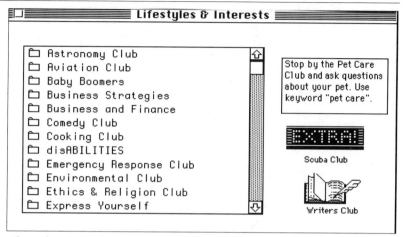

The *Lifestyles & Interests* department of America Online is the home of a number of special interest groups and clubs, including the Environmental Forum, the Astronomy Club, and the National Space Society.

Forum Organization

The Environmental Forum offers three types of resources: public discussion message boards, a software library, and a conference hall (see Figure 8-3). The discussion folders are broken down into four broad categories:

- The Front Desk (general environmental discussion, introductions of forum participants, and announcements)

- The Biosphere (issues that relate to air and water)

- Mother Earth (issues that relate to land and wildlife)

- Biotechnology Issues

Members can download and upload files, reports, and data to and from the software library, called the "Litter Free Library." And real-time interviews with environmental experts, general meetings, and lively debate and discussion can be found in the "Environ-Chat" conference room.

FIGURE 8-3

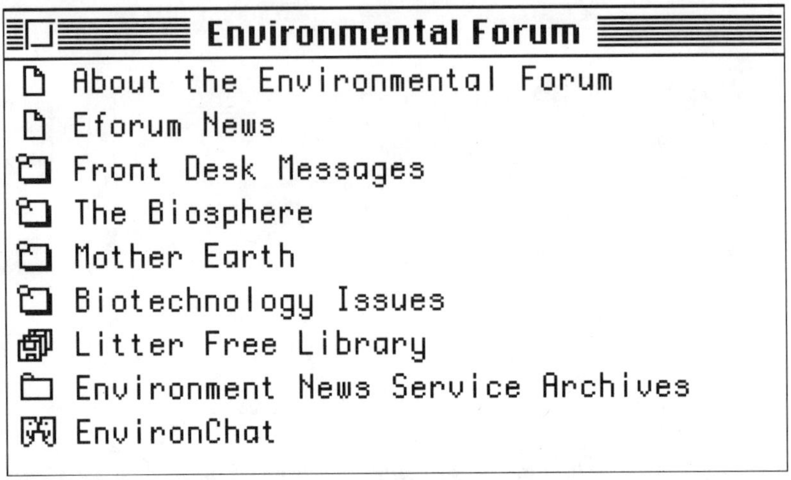

```
┌─────────────────────────────────────────┐
│ ▤□▤▤▤▤▤  Environmental Forum  ▤▤▤▤▤▤▤▤ │
│  ▢  About the Environmental Forum       │
│  ▢  Eforum News                         │
│  ▤  Front Desk Messages                 │
│  ▤  The Biosphere                       │
│  ▤  Mother Earth                        │
│  ▤  Biotechnology Issues                │
│  ▣  Litter Free Library                 │
│  ▢  Environment News Service Archives   │
│  ▩  EnvironChat                         │
└─────────────────────────────────────────┘
```

The main menu of the Environmental Forum shows the public discussion message boards, the software library, and the conference room for "live," real-time discussion of environmental issues. To enter any section, click on it with the mouse.

An Earthshaking Lesson

Teaching longitude and latitude to nine and ten year olds is not an easy task. Fourth-grade schoolteacher Linda Smith taught her students a lesson in real-life geology—and telecommunications—that they won't soon forget. She used a worldwide earthquake-tracking network provided by the U.S. Geological Service.

As part of a science fair project, Linda connected a personal computer in her Terre Haute, Indiana, classroom to a phone line. After a basic telecommunications lesson, she linked her students to the network.

"Each day, I would allow two students to access the earthquake-tracking network, and they would gather information on earthquakes that had happened in the past 24 hours. Those two students would then go to a very large map of the world, and using the longitude and latitude information given on the earthquake printout, they would use pins with colored heads to pinpoint each earthquake. It did not take long before the students began to see the various faults along which earthquakes occur."

Using a searchable encyclopedia on America Online, Linda's students looked up information about the population, the land formations, the climate, the soil type, and the known faults of each continent under study. By downloading the information into a word processor, the youngsters were able to prepare reports on known major faults, natural disasters following earthquakes, earthquake destruction, earlier major earthquakes of the 20th century, and earlier major recorded earthquakes.

"I was amazed at the geographical knowledge the students gained from this project. They really did enjoy themselves and were very eager to do the research to see why the earthquakes happened on some continents, and in some areas within a continent, more than on others. By continuously pinpointing the sites of earthquakes on the map, the students were easily able to learn the concepts of longitude and latitude."

Perhaps equally as important, the students developed the skills and self-confidence to use the vast electronic resources available for learning about the world.

"After a few days, they no longer needed my help. They could do it on their own!"

—Don Rittner

Public Message Boards

Forum participants can easily initiate, browse, or jump into public discussion on any environmental subject on the public messageboards (see Figure 8-4). Discussion topics are organized within each message board in "folders." Starting a public discussion is

FIGURE 8-4

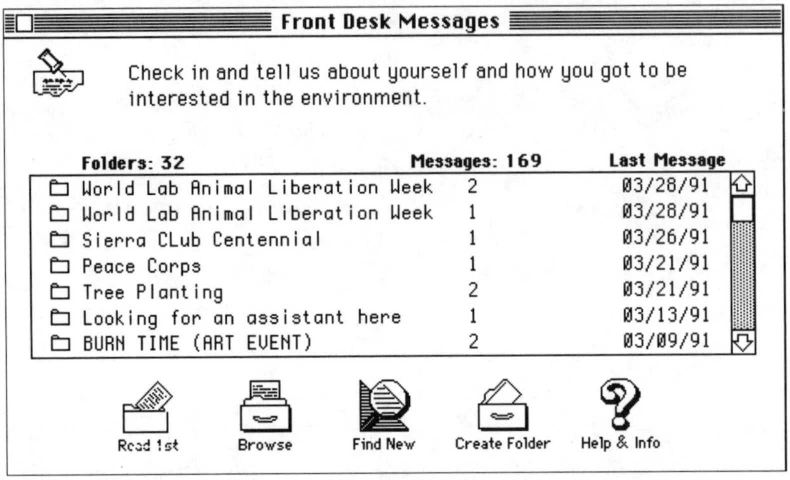

```
┌─────────────────── Front Desk Messages ───────────────────┐
│  📝  Check in and tell us about yourself and how you got to be │
│      interested in the environment.                            │
│                                                                │
│      Folders: 32              Messages: 169      Last Message  │
│      📁 World Lab Animal Liberation Week    2      03/28/91   │
│      📁 World Lab Animal Liberation Week    1      03/28/91   │
│      📁 Sierra CLub Centennial              1      03/26/91   │
│      📁 Peace Corps                         1      03/21/91   │
│      📁 Tree Planting                       2      03/21/91   │
│      📁 Looking for an assistant here       1      03/13/91   │
│      📁 BURN TIME (ART EVENT)               2      03/09/91   │
│                                                                │
│         Read 1st   Browse   Find New   Create Folder  Help & Info │
└────────────────────────────────────────────────────────────────┘
```

A list of public folders in the Front Desk public message board. The number to the right of each folder shows how many messages are in that folder, and the dates show when the folder was first created and last updated.

easy; you simply follow the graphic on-screen instructions to create a titled folder and post the first message, or you can comment on an existing topic, or add a message to the discussion (called a "thread") already going on in a folder of interest.

In the Front Desk message board pictured in Figure 8-4, for example, a forum member created a folder and typed the first message. Within a few days, he had an answer. Messages are organized chronologically within each folder. Readers can access messages in order of appearance, access one at a time, or access new messages since the last time they visited the folder, etc. (see Figures 8-5 and 8-6).

Software Library

In the Litter Free Library, members can upload or download files, reports, and data. The library allows quick access to documents; data can be downloaded in a matter of minutes (see Fig-

FIGURE 8-5

An opening message by a member requesting information about degradable plastics (see Figure 8.6).

FIGURE 8-6

The request in Figure 8.5 answered. Since America Online message boards are public, anyone can start discussions, respond to messages, or browse for information in the Forum.

ure 8-7). Most of the work of downloading and uploading has been built into the America Online software, so you don't need to remember complex commands. In other words, file transfer

FIGURE 8-7

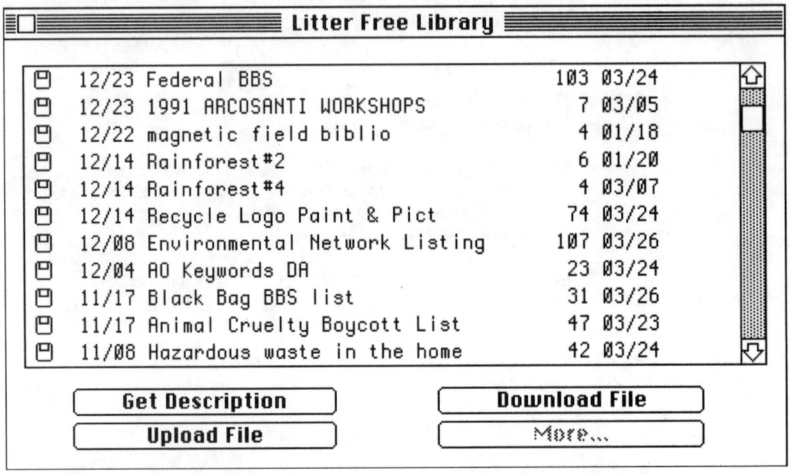

```
┌─────────────────────────────────────────────┐
│ ▤           Litter Free Library          ▤ │
├─────────────────────────────────────────────┤
│  📄  12/23 Federal BBS              103 03/24│
│  📄  12/23 1991 ARCOSANTI WORKSHOPS   7 03/05│
│  📄  12/22 magnetic field biblio      4 01/18│
│  📄  12/14 Rainforest#2               6 01/20│
│  📄  12/14 Rainforest#4               4 03/07│
│  📄  12/14 Recycle Logo Paint & Pict 74 03/24│
│  📄  12/08 Environmental Network Listing 107 03/26│
│  📄  12/04 AO Keywords DA            23 03/24│
│  📄  11/17 Black Bag BBS list        31 03/26│
│  📄  11/17 Animal Cruelty Boycott List 47 03/23│
│  📄  11/08 Hazardous waste in the home 42 03/24│
├─────────────────────────────────────────────┤
│      [ Get Description ]   [ Download File ]│
│      [   Upload File   ]   [    More...   ] │
└─────────────────────────────────────────────┘
```

A listing of some of the files available for downloading from the Environmental Forum software library. To transfer a file to your computer, select the file you want with your mouse and click on the Download File button.

protocols are transparent to the user. Use the mouse to click on the file you want, and it is downloaded to the "folder" or "directory" that you designate.

One of the many files that can be downloaded from the Litter Free Library is the Environmental News Weekly Reader. This HyperCard stack, designed for Macintosh users of the Environmental Forum by programmer Jeff Iverson, contains a week's worth of environmental articles taken from the various sources of America Online's News & Finance department. The stack gives environmentalists an easy way to keep track of environmental news.

Public and Private Conferences

America Online enables members to engage in "live," real-time discussions with each other in public and private conference rooms. This level of interaction allows forum members to participate—from their own home and office—in a lively, stimulat-

FIGURE 8-8

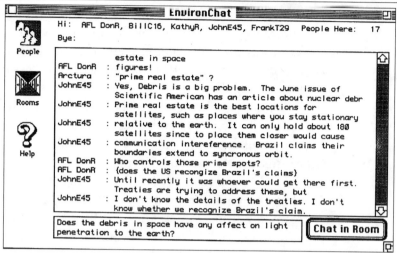

Express yourself! The EnvironChat conference room is the place for Forum participants to carry on real-time discussions.

ing debate and exchange of opinions that in the past could take place only in town hall and committee meetings. Figure 8-8 shows a group of Environmental Forum members discussing the environmental impact of space programs.

A number of special interest groups carry on weekly discussions in America Online public conference rooms. Members who prefer private discussions can create private "rooms" in the People Connection department of America Online.

Compton's Encyclopedia

Keyword: `encyclopedia`
Path: Learning & Reference > Compton's Encyclopedia

America Online's Learning & Reference department contains a valuable resource for environmentalists and general knowledge seekers of all ages—a searchable online encyclopedia. Unlike their printed counterparts, online encyclopedias are updated regularly and thus never go out of date. Articles can be searched by keywords and interest areas, and text can be saved for later reference or transferred into your word processing program for later manipulation.

Compton's Encyclopedia is published by Britannica Software, Inc., a division of Encyclopaedia Britannica, Inc. The encyclopedia features 8,784,000 words, 5200 full-length articles, 26,023 capsule articles, and 63,503 index entries.

The online encyclopedia is a fast and convenient way to check on environmental terms, theories, and concepts. Boolean searching (using AND, OR, and BUT delimiters) is available to add power and definition to your keyword searches. Clicking on the article titles displayed as a result of your search shows you the complete text (see Figure 8-9).

FIGURE 8-9

Look up environmental terms, theories, and concepts in an online, searchable encyclopedia. Subject search words and logical delimiters like AND, OR, and BUT help you narrow your search.

News Wire Services

> Keyword: news
> Path: News & Finance > Search News Articles > NewsWatch

In America Online's News & Finance department, you can find daily national and international news articles compiled from United Press International (UPI), Reuter's, Associated Press (AP), and other news services. NewsWatch allows you to search an up-to-date database of articles by keyword. Its search engine is similar to the encyclopedia's, including Boolean searching (see Figure 8-10). Text can be saved for later reference or manipulation.

FIGURE 8-10

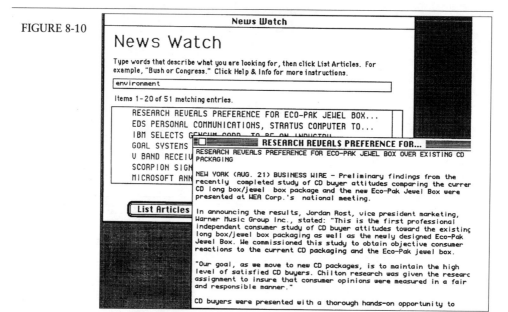

NewsWatch allows you to search for news that interests you by keyword. Delimiters like AND, OR, and BUT narrow your search.

Electronic Mail

> Path: Select Compose Mail from the Mail menu on the horizontal menu bar.

America Online's easy-to-use electronic mail system allows you to exchange private mail with fellow Environmental Forum members and other America Online members (see Figure 8-11).

The first time you sign on to America Online, you select a "screen name" that serves as your mail address and your "handle" in conferences or on message boards. Your screen name can be a combination of your first and last names or initials (DRittner) or a name that reflects your interests (Host Earth).

Once you have made initial contact with fellow members of the forum, you can keep track of them with your own online address book. You can address private mail simply by choosing the address book icon located on the Mail form.

You can also attach a file (for example, a word processing document, a spreadsheet, or a graphic) to your letters. Other mail

FIGURE 8-11

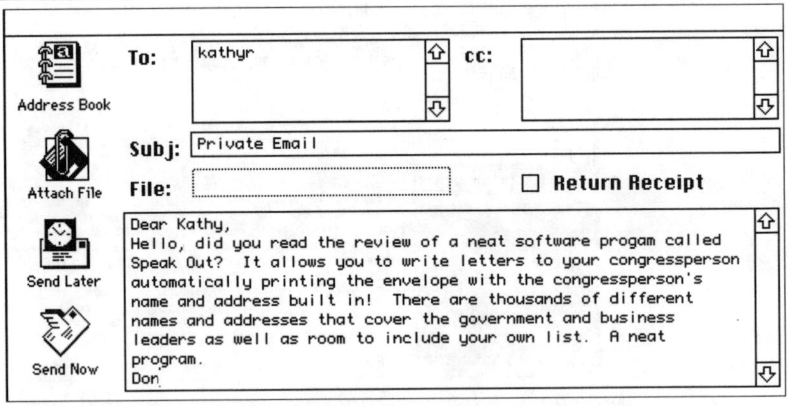

America Online's electronic mail system allows you to exchange private mail with other members. Since members can attach files to mail, important documents can be exchanged online.

options include sending group mail, forwarding mail, and sending "carbon copies" and "blind" carbon copies. With the Macintosh software version of America Online, you can schedule your computer to sign on automatically to send and retrieve mail. America Online also allows members to send U.S. mail and fax messages anywhere in the United States (no special equipment is needed).

> *The pressure of public opinion is like the pressure of the atmosphere; you can't see it, but, all the same, it is sixteen pounds to the square inch.*
>
> J.R. Lowell

CHAPTER 9

CompuServe

With over 500,000 members worldwide and over 1200 online offerings, CompuServe Information Service is the granddaddy of commercial online services. Started in 1969 as a time-sharing company, it went online to the public ten years later. Today, the "network nation" of CompuServe is one of the largest communities of personal computer users.

Since CompuServe is command driven, every major personal computer format is supported, including IBM PC, Macintosh, Apple II, Commodore, and Atari. If you are uncomfortable with terminal communications programs or would like offline information-processing capabilities, CompuServe offers graphic software "interface" programs, explained later.

CompuServe offers an extensive range of services, including special interest groups and clubs, computing support for virtually every type of computer, financial and stock information, and a broad range of reference databases. CompuServe electronic mail has links to MCI Mail, fax, telex, and the Internet.

How to Get Online

Anyone with a personal computer, a modem, and communications software can become a member of CompuServe. To get online, you need a CompuServe Membership Kit, which includes a user ID number and a temporary password to use the first time you sign on. CompuServe Membership Kits are available nationwide at almost any computer software retailer and at many bookstores. To find out where you can get one, call CompuServe customer service at 800/848-8199.

Cost

The CompuServe Membership Kit retails for $39.95 but includes $25 of connect time. Online charges vary based on your modem speed: 300 baud is $6.30 per hour; 1200 to 2400 baud is $12.80 per hour; and 9600 baud is $22.80 per hour. You pay a $2 online support fee every month to maintain your subscription to CompuServe.

In addition, members are charged network communications surcharges for use of SprintNet and Tymnet telephone numbers for each session on CompuServe. Some features, particularly reference and financial databases, carry a premium surcharge.

Organization

CompuServe services are divided into ten major subject areas (see Figure 9-1):

- Communications/Bulletin Boards
- News/Weather/Sports
- Travel
- The Electronic Mall/Shopping
- Money Matters/Markets

FIGURE 9-1

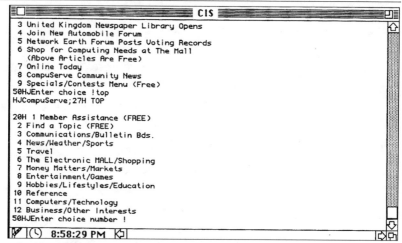

The top-level menu of CompuServe (as accessed through the Macintosh version of MicroPhone communications software). To go into an area that interests you, type the menu item number after the ! prompt.

- ◆ Entertainment/Games
- ◆ Home/Health/Education
- ◆ Reference
- ◆ Computers/Technology
- ◆ Business/Other Interests

Getting Around

You may navigate through CompuServe using one of three methods:

- ◆ CompuServe offers a system of menus; by selecting an item off each one you come to, you "branch down" from top-level menus into submenus.

- ◆ Together, GO commands and "quick reference words" allow you to "jump" immediately to the service you are looking for, without branching through menus and submenus. You can type a GO command whenever you see a ! prompt on the screen. Type `GO` and the quick reference word of the area you are interested in—for example, `GO EARTH`. Additional commands to help you navigate CompuServe menus are listed in Figure 9-2.

♦ CompuServe offers a software program called CompuServe Information Manager that acts as a graphic "front end" for the service. Information Manager, available for IBM PC compatibles and the Macintosh, uses the windows, pull-down menus, and icons that are familiar to Mac users and becoming increasingly familiar to PC users (see Figure 9-3). Information Manager makes it easy to navigate through CompuServe, since you don't have to memorize commands. You can find Information Manager at many software retailers, and you can order it online by typing `GO ORDER` at the ! prompt.

FIGURE 9-2

`T`		**Top.** Takes you to the top-level menu of CompuServe.
`M`		**Menu.** Takes you back to the previous menu.
`? or H`		**Help.** Displays a list of the commands that you can use.
`FIND xxxx`		**Find.** Typing `FIND` and the name of a topic displays a list of quick reference words for that topic.
`N`		**Next.** Displays the next item off the last menu that appeared on your screen.
`P`		**Previous.** Displays the previous item off the last menu that appeared on your screen.

These keyboard commands can help you navigate through CompuServe's menu. Commands can be typed after the : or ! prompt. After typing a command, press Return.

FIGURE 9-3

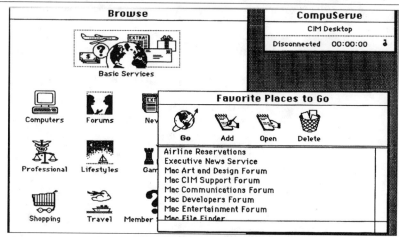

CompuServe Information Manager (CIM) software offers an easy-to-use, graphic "front end" for people who are uncomfortable with a text command-driven system. This screen shows Information Manager for the Macintosh; an IBM PC-compatible version is available as well.

Since most people who use CompuServe use the service's text-based, command-driven mode, this chapter references GO commands for CompuServe services.

Environmental Resources

CompuServe offers a great depth and breadth of resources covering virtually every area of interest. This chapter explores the following CompuServe resources which are of particular interest to EcoLinkers:

- The Network Earth Forum
- The Science and Math Educational Forum
- The Good Earth Forum
- The Outdoor Forum
- The SafetyNet Forum

Introduction to CompuServe Forum Organization

Every CompuServe forum contains the following elements:

- an announcement area,
- a software library,
- a message board,
- a conference area, and
- a member directory.

CompuServe forums are run by forum administrators, who are responsible for maintaining the message boards and libraries and running forum events.

File Libraries

Each file library in a CompuServe forum contains software and articles related to the special interest of that forum. Figure 9-4 shows the top level of the Network Earth Forum. To enter the

FIGURE 9-4

```
HOUSE.LCV and SENATE.LCV in the
Legislation library. These files
contain the Environmental Scorecard
developed by the League of Conservation
Voters.

Remember, NETWORK EARTH airs at 11:00
p.m. Eastern time, 8:00 p.m. Pacific
time on TBS!

Press <CR> !
HJNetwork Earth Forum;30HForum Menu

20H 1 INSTRUCTIONS

   2 MESSAGES
   3 LIBRARIES (Files)
   4 CONFERENCING (0 participating)

   5 ANNOUNCEMENTS from sysop
   6 MEMBER directory
   7 OPTIONS for this forum
50HJEnter choice !
```

The menu of the Network Earth Forum (GO EARTH). To enter any area of the forum, simply type the number of the menu item after the ! prompt.

file library, you type its number—3. You are then presented with a list of the library subtopics (see Figure 9-5). After you have chosen a subtopic, you are offered options that let you browse through descriptions of available files or examine a directory of files without descriptions.

When you have selected a file you want to download, CompuServe will prompt you to select a file-transfer protocol. Be sure to select a protocol supported by the communications program you use to access CompuServe. Your communications program prompts you to name the file you are transferring to your computer and to indicate the directory on your computer to which it should be transferred. CompuServe notifies you when the file transfer is complete.

Message Boards

On message boards, you can participate in discussions with other forum members. If you select item 2 on the Network Earth menu shown in Figure 9-4, you are presented with the menu shown in Figure 9-6.

FIGURE 9-5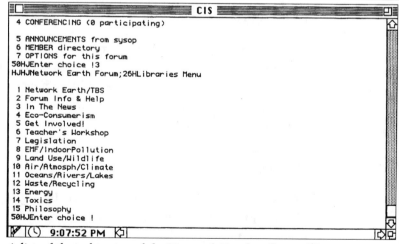

A list of the subtopics of the Network Earth software library.

FIGURE 9-6

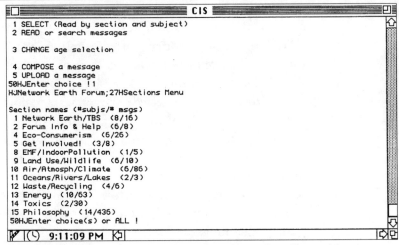

The message board menu of the Network Earth software library.

CompuServe allows you to look through messages that interest you by subject matter or in chronological order. Once you have selected a message that interests you, you can follow the "thread" of replies to it or add a message of your own.

Conference Areas

In CompuServe's live conference areas, you can participate in formal and informal real-time discussions with other forum members. Many CompuServe forums have weekly or monthly meetings. You can find conference schedules by selecting the Announcements menu item of each forum.

Member Directory

The first time you enter a CompuServe forum, the system prompts you to "join" the forum by entering information about yourself in the forum Member Directory. This database can be searched by user ID, name, and interest. It's a great way to find other EcoLinkers who share your interests.

Network Earth: An Online Experiment in "Participatory Television"

An exciting new form of environmental news debuted on national cable television in August 1990, when Network Earth—a half-hour environmental news show—was aired around the country. The show is Turner Broadcasting System's attempt to bring environmentally aware programming to the public, with a twist: viewers who have a personal computer and a subscription to CompuServe can interact immediately after the show with the producers, environmental experts, and other online users. This interaction allows CompuServe members to talk about applying the environmental solutions they've learned from the television show to their own communities.

The Network Earth Forum is the online complement of the show. This is how it works: After watching Network Earth on TV, simply sign on to CompuServe, and type `GO EARTH` at the ! prompt. You can immediately discuss the show with other viewers, give feedback to the producers, or access more information on the subject area covered in the show. You can even download a transcript of the show if you missed it. The libraries offer files with tips on how to lead an environmentally sound lifestyle.

As a member of the Network Earth Forum, you can suggest story ideas to the show's producers and, if you have a video camera, produce your own environmental "short" that may be aired on the show.

The premiere show featured stories on the environmental impact of the 1989 Panama invasion; the Baywatchers, a local ecodefense force patrolling San Francisco Bay to protect it from polluters; science writer Roger Bingham's excess packaging piece; and an interview with environmentalist Michael Stipe, the lead singer of the rock band REM.

According to associate producer Steffan Sandberg, the administrator of the Network Earth Forum, participatory television shows great promise. More innovative programs such as this, combining the communications power of television and the home computer in an easy to digest format, will bring the average viewer closer to understanding global environmental issues, as well as broaden national awareness of and interest in EcoLinking.

The Network Earth Forum also offers more than a dozen message boards and file sections, including information on how to get involved in the environmental movement, contacts in and addresses of environmental organizations, what you can do at home or in the workplace, tips for projects in your community, global issues, projects for kids, and a teachers' workshop for educational idea exchange.

The Science/Math Educational Forum

The Science-Math Educational Forum can be accessed by entering the command GO SCIENCE. This very active CompuServe forum has nearly 7000 members, ranging from students and teachers to doctoral-level astrophysicists. Rick Needham, Chair of the science department at Mercersburg Academy, Mercersburg, Pennsylvania is the forum administrator. File libraries contain hundreds of software programs for the home and the classroom, covering a wide range of science and math topics.

Forum features include

- late-breaking news from many of the leading science magazines;

- the Lab Notebook from which you can learn safe experiments;

- listings of science and math job opportunities;

- an area in which students can practice for college boards; and

- information on science and math fellowships, scholarships, and assistantships for college students.

The conference area of the forum hosts weekly real-time conferences for math and science teachers, as well as weekly exchanges among scientists and science educators. To find out more about conference schedules, check the announcements area of the forum.

The Good Earth Forum

The Good Earth Forum—accessed by entering GO GOODEARTH —is devoted to folks interested in the back-to-basics movement. Topics include lifestyles, gardening, houseplants, nutrition, folkways, pets, and herbs. Newspaper columnist and writer Dave Peyton is the forum administrator.

Through the Good Earth Forum, you can make contacts to collect and exchange plant seeds, get your questions answered about vegetable gardening, learn ornamental horticulture and landscaping, learn the basics about herbs, discuss nutrition and ecology, and explore how other cultures work the "natural" way. You can also download files full of information that will help you enjoy the good earth.

The Outdoor Forum

The Outdoor Forum is a great way to exchange information with outdoor lovers around the country. Topics of discussion include fishing, hunting, camping, scouting, cycling, birding, climbing, boating, skiing, photography, and other outdoor activities.

Outdoor Forum message boards and file libraries—accessed by entering GO OUTDOOR FORUM—span a wide range of topics, including

- Scouting
- Outdoor Photography
- Trout Unlimited
- Fishing
- Hunting
- Cycle/Run/Walk
- Wildlife/Birding
- Boat/Water Sports
- Camping/RV
- Snow Sports/Climb
- OWAA (Outdoor Writers Association of America)
- Firearms
- National Rifle Association
- Environment
- *Outdoor Life* magazine

The Outdoor Forum is also the electronic home of the Outdoor Association of America. In an Outdoor News Clips area, articles from AP, UPI, and others about outdoor topics are posted regularly.

The forum administrators are Joe Reynolds, the northwest regional editor of *Field and Stream* magazine; Bill Clede, former outdoor editor of the Hartford (Connecticut) *Times* and environmental director for Hartford's WTIC radio; Les Line, editor-in-chief for 25 years of *Audubon* magazine; and Tony Mandile, Arizona editor of *Outdoor Life* magazine.

The SafetyNet Forum

The SafetyNet Forum provides information on all areas of safety. Forum members include professionals in occupational health, safety engineering, and fire prevention and others interested in safety issues. Enter `GO SAFETYNET` to access this forum.

Hosted by industrial hygiene consultant Charles M. Baldeck, SafteyNet has message boards and file libraries covering topics such as:

- chemical and physical hazards
- biohazards and radiation
- waste management
- general environmental issues
- fire and emergency medical services
- transportation issues
- consumer products and safety
- police business and emergency planning

Electronic Mail

In addition to letting you correspond with other members, CompuServe's electronic mail system—Easyplex—allows you to exchange mail with users of MCI Mail and the Internet. You are also able to send correspondence by fax and U.S. Mail and to send and receive messages with telex machines worldwide. Type `GO MAIL` to access CompuServe's mail features.

Other Useful Resources

Many databases that may be of use to you in your pursuit of information on the environment are available on CompuServe. They include:

- IQuest
- NORD Services/Rare Disease Database
- PaperChase/MEDLINE
- Physician's Data Query
- Books in Print
- CENDATA/The Census Bureau Service
- The National Technical Information Service

Although many of the above can be accessed through database services such as Dialog Information Service (described in chapter 13), you can find them on CompuServe as well (albeit generally for an additional surcharge).

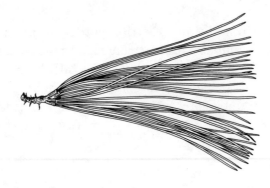

What is the use of a house if you haven't got a tolerable planet to put it on?

Henry David Thoreau

CHAPTER 10

EcoNet

EcoNet ("the environmental network"), founded in 1987, is an essential resource for individuals and organizations that want to stay on top of worldwide environmental issues. EcoNet's single greatest contribution is that it enables its members to stay in touch with fellow environmentalists in over 70 countries, using dozens of e-mail gateways.

EcoNet is part of the Institute for Global Communication, a San Francisco organization that is composed of several networks working together, including

- PeaceNet
- EcoNet
- HomeoNet
- ConflictNet
- Alternex (Brazil)
- FredsNaetet (PeaceNet Sweden)
- GreenNet (England)
- Nicarao (Nicaragua)

- Pegasus (Australia)
- The Web (Canada)

Because EcoNet works in conjunction with the Association for Progressive Communications (APC), EcoNet members can also communicate with the networks participating in that organization.

Hundreds of environmentally concerned organizations and individuals use EcoNet in a variety of ways. EcoNet members arrange local, regional, national, and international conferences. Environmental groups regularly post alerts requesting letter-writing campaigns and information. Environmental organizations post electronic newsletters for downloading or reading online. Other organizations download posted articles for their own newsletters.

Grant information is available online, and you can read press and news releases. An online version of the National Wildlife Federation's Conservation Directory lists virtually every environmental organization in North America. You can also find a complete list of U.S. Congresspeople and their membership on various environmental committees.

Frequent news contributors include the Sierra Club, Friends of the Earth, the Center for Conservation Biology at Stanford, the International Union for the Conservation of Nature and Greenpeace. The Environment News Service provides constant updates on worldwide environmental news.

EcoNet is more than just an environmental news source, however. EcoNet is a worldwide environmental conference center, offering more than 600 conferences covering virtually every environmental topic. You can meet friends on EcoNet who share a common environmental concern, and you can form work groups or alliances to work on environmental issues that interest you.

How to Get Online

Contact the Institute for Global Communications at 415/442-0220 for membership information. Or write to Institute for Global Communications, 18 de Boom, San Francisco, CA 94107.

Cost

A one-time sign-up fee of $15 includes an excellent 150-page manual and one hour of online time. A monthly membership fee of $10 includes one free hour of off-peak time per month. After that, you pay $5 per hour for off-peak use (6 p.m. to 7 a.m. weekdays, all days on weekends and selected holidays) and $10 per hour for use during peak time.

Members are allotted 100K (about 50 pages) of free space for storage of e-mail or other material on the EcoNet system; a surcharge of one cent per page per month is charged over that. If you refer new members to EcoNet, you are rewarded with free online time.

Getting Around

EcoNet is a command-driven system. When you first sign on, you are presented with a main menu listing five options. Typing the first letter of an option at the ? prompt lets you gain access to that function of EcoNet. EcoNet commands and their functions include the following:

(c)onf takes you into the many conferences.

(h)elp gets you extensive online help.

(m)ail lets you send private electronic mail.

(s)etup lets you change your password, set the terminal type, and perform other functions.

(u)sers gives you a list of users on the various networks that comprise EcoNet.

(b)ye logs you off of EcoNet.

If you ever get lost on EcoNet, typing q for *quit* at the ? prompt takes you back to the main menu.

Environmental Resources

EcoNet's scope encompasses the broadest possible definition of environmental concerns, including the economic, sociological, and political aspects of the discipline. This chapter explores the following important EcoNet resources:

- Access to environmental networks and associations
- Electronic mail gateways
- Environmental conferences

Environmental Networks and Associations

Close to 200 international networks and associations support conferences or online discussion groups on EcoNet. To get a current list of online organizations, type c (for conference) at the ? prompt on the main menu, and then enter the command shown in boldface in Figure 10-1.

Table 10-1 lists a sample of the networks and organizations that you can reach through EcoNet. Once you locate the acronym for the organization or network you are interested in, you can find all the conferences supported on EcoNet by that organization or network by typing the acronym when prompted on the conference menu.

FIGURE 10-1
```
-- conference directory --
Network: (a)ll, (o)ther, or <RETURN> for EcoNet (? for
help): o
Other: (o)rg, (u)senet (? for help): o
Organizational acronym ('l' to (l)ist all, or ? for help): l
Org list: (a)ll, (o)ther, or <RETURN> for EcoNet (? for
help): a
```

Enter the choices shown in boldface to get a list of organizations currently sponsoring conferences on EcoNet.

TABLE 10-1
A selection of the almost 200 international organizations that support conferences on EcoNet.

Acronym	Organization
AAN	Anti Apartheid Network
ACD	Alliance for Cultural Democracy
AI	Amnesty International
APC	Association for Progressive Communications
APT	American Peace Test
AWEA	American Wind Energy Association
BW	Beyond War Foundation
CAB	Council for All Beings
CARC	Central America Resource Center
CARNet	Central America Resource Network (CARNet)
CC	Common Cause
CCB	Center for Conservation Biology
CDC	Centers for Disease Control/AIDS Weekly InfoLine
CDI	Citizen Diplomacy, Inc.
CEC	Center for Economic Conversion
CI	Christic Institute
CID	Center for Innovative Diplomacy
CIIS	California Institute of Integral Studies
CLW	Council for a Livable World
CND	Campaign for Nuclear Disarmament
CSIA	Canadian Social Investment Association
CWC	California Wilderness Coalition
DN	Dolphin Network
ED	Earth Day
EF!	Earth First!
EGA	Environmental Grantmakers Association
EIN	Earthcare Interfaith Network
FCNL	Friends Committee on National Legislation
FN	FarmNet
FOE	Friends of the Earth
GAIA	Green Alternative Information for Action
GAN	Global Action Network
GCC	U.S. Greens
GN	GreenNet
GP	Greenpeace

Acronym	Organization
GPM	Great Peace March
GPOC	Green Party of California
GREENS	Greens
GRI	Grassroots International
GWLW	Global Walk for a Livable World
IBASE	IBASE/Alternex
IBN	International Business Network
ICA	Institute of Cultural Affairs
IGC	Institute for Global Communications
IPPNW	International Physicians for the Prevention of Nuclear War
ISCOS	Institute for Security and Cooperation in Outer Space
IUCN	International Union for the Conservation of Nature
LSR	Lawyers for Social Responsibility
NAEE	National Association for Environmental Education
NAN	North Atlantic Network
NCC	News on Citizen Computing
ND	New Directions
NEC	Northcoast Environmental Center
NEFA	Northeast Forest Alliance
NEP	Nicaraguan Entrepreneur Project
NFD	Network for Democracy
NGLS	United Nations Nongovernmental Liaison Service
NLG	National Lawyers Guild
NMS	National Mobilization for Survival
NPC	National Peace Council
NRCC	Northern Rockies Conservation Cooperative
NUSP	Options 2000 Nuclear Sovereignty Project
NW	Nuke Watch
NWAG	New World Agriculture Group
NWF	National Wildlife Federation
NWFC	Nuclear Weapons Freeze Campaign
POPTEL	GeoNet Bulletin Board
PSR	Physicians for Social Responsibility

Acronym	Organization
RAN	Rainforest Action Network
RIC	Rainforest Information Center, Australia
SC	Sierra Club
SCI	Service Civil International
SF	SANE/Freeze
SI	Servas International
TECNICA	Institute for Technology and Development
TWR	Third World Resources
UCS	Union of Concerned Scientists
UDC	Union for Democratic Communications
UN	United Nations
UNA	United Nations Association
UNCED	U.N. Conference on the Environment and Development
UNDP	United Nations Development Programme
UWC	United World Colleges
VANA	Veterans against Nuclear Arms
WAWF	World Association for World Federation
WCN	World Citizens Network
WEB	Web, Canada
WRM	World Rainforest Movement
WS	Windstar Foundation

Electronic Mail Gateways

One of the greatest tools on EcoNet is the ability to send mail to people around the world. You can send electronic mail to anyone who has an account on one of the either noncommercial or commercial networks that are gatewayed through EcoNet. To send mail, type m at the ? prompt on the EcoNet main menu. All you need to know are the acronym of the network to which the mail recipient belongs and the account name or user ID the recipient uses on that network.

Table 10-2 shows how to address mail to the commercial networks that can be reached through EcoNet (at a small additional fee). Table 10-3 shows how to address mail to the noncommercial networks that can be reached through EcoNet (at no additional fee).

TABLE 10-2 Commercial Network Addresses	Network	Address
	Alternex	ax:<account>
	AppleLink	apple:<account>
	ATT E-Mail	attmail:<account>
	BIX	bix:<account>
	ClariNet	clarinet:<account>
	CGNet	cgnet:<account>
	Connect	connect:<account>
	DASNET	dasnet:<account>
	Dialcom	dialcom:<host>:<account>
	EasyLink	easylink:<account>
	EIES	eies:<account>
	Fax	fax:cccnnnnnnn (c is country code, c is area code, and n is phone number)
	FidoNet	<account>@f###.n###.z#fidonet.org
	FredsNaetet	pns:<account>
	GreenNet	gn:<account>
	Handsnet	handsnet:<account>
	IGC Networks	<account>
	IM	imc:<account>
	INET	inet:<account>
	Janet	janet:<account>@<host>
	MCI E-Mail	mci:<account>
	Metanet	meta:<account>
	Nicarao	ni:<account>
	NWI	nwi:<account>
	Pegasus Networks	peg:<account>
	SF-Moscow Teleport	sfmt:<account>
	TCN	tcn:<account>

Network	Address
Telemail (Telecom Canada)	<telemail system>:<account>
Telex	telex:<country code><telex number>
The WELL	well:<account>
TWICS BeeLine (Japan)	twics:<account>
UNISON	unison:<account>

TABLE 10-3 Non-commercial Network Addresses

Network	Address
BITNET	<account>@<host>.bitnet
CSNet	<account>@<host>.csnet
Internet	<account>@<host>.org.type
Portal	portal:<account>
UUCP Mail Net	<account>@<host>.uucp

EcoNet Conferences

EcoNet carries hundreds of conferences that fall into over 50 broad categories, called topics, that cover all aspects of the environment and sciences. In this section, you will learn how to find conferences that interest you and how to participate in them. A sample listing of EcoNet conferences follows.

Finding a Conference That Interests You

To find conferences on EcoNet, go back to the EcoNet main menu and type c at the ? prompt. You are presented with a list of over 50 categories of conferences, each with a number. By typing the number of the category that interests you, you receive a list of the conferences in that category. To enter a conference, type the conference ID at the ? prompt.

The Global Office

Sanford Lewis, an environmental attorney, works for the National Toxics Campaign Fund (NTCF) in Massachusetts. Lewis and his colleagues have written reports for Congress and the national media, such as "Shadow on the Land," about farm chemicals, and "From Poison to Prevention," a critique of EPA's incineration versus waste reduction policies. Lewis's work requires him to fly around the United States a lot. Much of his work is done in airplanes, editing documents on his Toshiba laptop computer.

The NTCF uses e-mail to exchange policy ideas among colleagues located in widely dispersed offices. Each member has a different area of expertise. He or she will work on a document, then pass it on via e-mail to the next person.

In August, 1989, NTCF gave simultaneous press conferences in 40 states on the release of the "Poison to Prevention" report. Greenpeace put a simultaneous notice on the press conference in its newsletter, and EPA offices were blitzed with phone calls.

"We were putting pressure on them," Lewis said. "They made some adjustment in their policy, but not enough that we don't have to continue working on this in a big way."

Every two or three months, Lewis gets an inquiry from another country. "People have found me through my postings, and they ask, 'Do you know anything about this?' One of the more interesting ones was in 1990 with a group in South Africa, Earthlife. They had a meeting with some oil companies coming up in two days." I explained what the National Toxics Campaign Fund is doing around petrochemical facilities; we've been trying to get them to enter into Good Neighbor agreements, allow citizens to inspect them, and negotiate for pollution prevention. I immediately sent them a compilation of materials on this."

A week later, Lewis got a message back from Earthlife saying that they liked his message and that they were adopting the Good Neighbor strategy.

He has since sent the South African group a list of requests for information NTCF had submitted to Exxon in Texas, and described some problems NTCF had been having with Chevron in Richmond, California.

"Chevron does business in South Africa. We wanted to get Earthlife to work on the Chevron affiliate there, so we could join forces and pressure them from both places," Lewis said.

Lewis can be reached through EcoNet's user directory.

—Wendy Monroe

Participating in Conferences

EcoNet conferences are divided into topics, each with responses posted by EcoNet members. You can read through a conference by typing u at the ? prompt to read the first unread message in the earliest unread topic.

To save time and money you can download the topics by typing c to go into capture mode. You can use either ASCII (text), Kermit, or Xmodem protocol.

To respond to a topic, type W to write. You have the option of sending your response to the conference or replying privately to the original author. You can also write a new topic by typing wnc (writing new message to the conference). Many similar commands that can help you enjoy conferencing on EcoNet are explained in the user manual.

A Selection of EcoNet Conferences

The following is a representative sample of conferences available to members of EcoNet. You can go directly to a conference by typing its ID at the conference prompt (IDs are the names shown in boldface in the following entries).

Air and Climate

en.cleanair Communication about air pollution; topics range from government policy to action ideas.

en.climate Discussion of pollution, its effects, and methods for dealing with it.

en.pollution Conference on pollution, including effects, sources, policy, and advocacy.

en.waste Discussion of waste management, from toxic pollution to actions planned against polluters.

gan.acidrain Comprehensive information about acid rain and what you can do about it.

gan.globalwarm Comprehensive information about global warming and what you can do about it.

Alerts

en.alerts Short and urgent announcements and alerts for the environmental movement.

gan.actionaler List and status of current actions related to a variety of environmental issues.

Announcements

en.announcemen Announcements, news, and notices of events pertaining to environmental issues.

en.jobs Job listings in the environmental field (mainly in the U.S.).

Beyond War

bw.environ Forum to exchange ideas, information, and resources in an attempt to address the question of how to live in harmony with the planet.

bw.planet Communication on Beyond War's Earth in Every Classroom project.

Calendars

en.calendar Announcements of environmental rallies, conferences, and other events.

Development

at.general Worldwide news and networking about the design and implementation of appropriate solutions to ecological and international development problems.

gan.population Comprehensive information about population issues and what you can do about them.

ppn.food+hunge Discussion and information concerning food (quality, contamination, nutrition, agriculture) and world hunger (causes, occurrences, solutions, policies).

East-West

en.ussr Up-to-date forum on the growing numbers of U.S. and Soviet NGO activists pursuing the goal of global environment sustainability with an emphasis on the protection of Soviet ecosystems.

E-LAW's Secret Weapon

John Bonine, a law professor at the University of Oregon in Eugene, heads the United States office of the Environmental Law Alliance Worldwide (E-LAW), founded in May, 1991. In the first six months of the organization's existence, U.S. E-LAW's staff of four has provided research and active support to local lawyers in Japan, Uruguay, Chile, Panama, Nicaragua, and Russia.

E-LAW has already established offices in Sri Lanka, Malaysia, Indonesia, Australia, The Philippines, Ecuador, Peru, and the United States.

E-LAW's mission is to provide research support to foreign public interest lawyers. E-LAW's North American office has the cheapest access to legal information and often receives urgent requests for legal research or expert testimony in a case. Staffers scramble to find a legal precedent, or locate the right expert witness. Bonine stresses that E-LAW's goal is to support local lawyers in their work, not to do the work for them. "We are simply their long-distance eyes and ears."

E-mail is E-LAW's secret weapon against communication costs. Not only is e-mail far cheaper than courier services and faxes, it's also an ideal medium for complex documents to be divided up by several people and worked on simultaneously.

Bonine feels that environmentalists have no choice but to become sophisticated in computer telecommunications as their corporate adversaries. "We're facing a well-organized and integrated effort by multinational companies and their law firms," Bonine says. "They have the resources to look around and pick on the countries most vulnerable to destructive development. If we can't catch up with them, we'll fall behind."

Sometimes E-LAW's research scores a direct hit, such as the time it sent a quick infusion of EPA studies on the health effects of high-voltage power lines to lawyers arguing before the Uruguayan parliament. The parliament, then deliberating on whether to destroy some centuries-old trees in the construction of power lines through a poor neighborhood, voted to suspend the project immediately.

E-LAW's biggest victory to date may have established a precedent that will keep Australian logging companies on the defensive for quite some time.

Tim Robertson, a lawyer in Sydney, Australia, had brought suit in August, 1991 before the Land and Environment Court of New South Wales. His case charged that logging in Chaelundi Forest would jeopardize the biological diversity of the area and the habitat of 23 endangered or protected species, including the Sooty Owl. He needed to establish that logging the forest would cause a decline in the species, and his argument would be shaky unless he received evidence within 24 hours.

Robertson needed the testimony of an expert, but he had no luck finding one in Australia. He called the E-LAW Australia office, which sent an urgent message to the E-LAW U.S. office in Eugene by fax and electronic mail.

By chance, an American expert on the Northern Spotted Owl had recently returned from a visit to Australia to study the Sooty Owl. E-LAW was able to reach him, and his testimony was in Robertson's hands in Sydney within the 24-hour deadline.

This became the essential evidence that convinced Presiding Justice Paul Stein to order the logging suspended. In late September, the Court declared that logging in part of Chaelundi State Forest would be in breach of the prohibition in the National Parks and Wildlife Act on taking or killing endangered species.

E-LAW's success may seem amazing for a brand-new organization, but Bonine refuses to rest on his laurels. As he puts it, "In environmental matters, all victories are temporary, and all

> defeats are permanent. We can only string together temporary victories to a semblance of permanency."
>
> Meanwhile, news of the victory took on a life of its own. E-Law's offices produced a press release on the logging suspension and posted it in several EcoNet conferences. EcoNet's administration permitted E-LAW to post a brief "banner" on the Sooty Owl victory that would automatically come onscreen for anyone logging onto EcoNet. Sunny Lewis of the Environment News Service in Vancouver, Canada, saw the story and ran it, and thus it was faxed and e-mailed around the world.
>
> John Bonine and E-LAW can be reached on EcoNet in the elaw.public interest conference.
>
> —Wendy Monroe

whatsnext Often a digest of other APC conferences, the focus of this one is on new directions in East-West relations and how to increase links among the environment, social justice, and economic development.

Education and Research

ccb.update Newsletter from the Center for Conservation Biology at Stanford University.

Energy

awea.windnews Discussion of wind energy, including back issues of the American Wind Energy Association newsletter.

en.bioanaerobi Discussion of alternative methods of recycling waste products for fuel.

en.bioconversi Discussion of the use of woody and agricultural plants in energy production.

en.energy Discussion, news, and requests on all aspects of energy.

en.toxics.inci Conference on hazardous waste incinerators, and waste-to-energy plants.

gn.nuclear Information on and discussion of campaigns against the nuclear industry in Europe. Some topics are in German, but the facilitator provides abstracts in English.

ppn.nukemateri Discussion and information about the nuclear fuel cycle, from mining through waste handling and its effects on the land and people.

sci.energy A Usenet conference on energy. Networked with thousands of academic institutions.

transport Discussion of transport issues, ideas, discoveries, inventions.

ws.connections General information about Windstar Connections, their role in the Windstar Foundation, and how organizations can "think globally but act locally."

ws.media Newsletters published by local Windstar Connections, as well as articles and print materials of interest to those dedicated to creating a sustainable future.

ws.networking Information about people, places, and tools for making the work of creating global sustainability more smooth, effortless, and fun.

Education

akashic.record Resources, culled from various print sources, to use in debates and education relevant to change and the evolution of a saner world.

aee.biblio Teaching aids, classroom experimenting and exploring kits, books on environmental education methods, and topics to cover. A conference of the Alliance for Environmental Education.

aee.curriculum Reports on trends and practices found in current environmental education curricula. A conference of the Alliance for Environmental Education.

ed.campusearth Campus Earth is a networking conference for college students, who exchange information on environmental news, issues, projects, ideas, and contacts.

en.enveducatio List of selected publications, events, and awards of interest to environmental educators.

naee.bibliogra Directory of resources for environmental education.

naee.databases List of environmental education databases and other searchable electronic resources.

nwf.consdirectory The entire text of the 35th edition of the Conservation Directory.

Environment

bw.environ Forum to exchange ideas, information, and resources that help address the question of how to live in harmony with the planet.

ega.directory The Environmental Grantmakers Association Directory is a working document of approximately 90 foundations interested in environmental grantmaking.

ega.grants Quarterly reports of grants made by foundation members of the Environmental Grantmakers Association.

en.general General discussion of the environment and the impact of humankind.

en.user Up-to-date forum on the growing numbers of U.S. and Soviet NGO activists pursuing the goal of global environment sustainability with an emphasis on the protection of Soviet ecosystems.

gan.actionguid A step-by-step guide to starting an action group or strengthening an existing organization.

gan.agencies Addresses and phone numbers of key U.S. Government offices and agencies engaged in environmental matters.

gan.housecommi List of House of Representatives committees dealing with specific environmental issues.

gp.press News service from Greenpeace that carries European stories.

gpty.general Discussion of the Green Party and green politics.

green.genetech News and discussion about the campaign to prevent genetically engineered products from being introduced into the environment.

intl.volunteer International volunteer work for peace and the environment.

iucn.news Species conservation, tropical forests, critical sites, and populated areas, wetlands, population, environmental education, sustainable development, and computerized environmental monitoring.

oz.attitudes Examines attitudes that hinder peace and environmental reforms.

oz.ecofeminism Discussion of an evolving feminist approach to ecology and earth-relatedness.

sol.news News digests and editorial commentary from *Solstice* magazine, the Journal of Personal and Planetary Health. *Solstice* emphasizes personal responses to ecological challenges.

talk.environme A Usenet conference on the environment. Networked with thousands of academic institutions in the United States and Europe.

ucs.updates Updates about projects and activities of the Union of Concerned Scientists.

Food and Agriculture

en.agriculture Discussion and news on current trends in alternative agriculture, rural sociology, agroeconomics, and the politics of agriculture.

en.toxics.pest Conference on insecticides, herbicides, fungicides, rodenticides, and other pesticides.

Forests

en.caforest Information on California forests.

en.forestplan Discussion and information on forests and their protection.

en.parks Information and discussion on national and other parks, and their management.

gan.tropicalfo Comprehensive information about tropical rain forests and what you can do to help preserve them.

ran.ragforum Facilitates the gathering and dissemination of information for Rainforest Action Groups throughout the world.

ric.wrr Online version of the *World Rainforest Report* published by the Rainforest Information Centre of Lismore, New South Wales, Australia.

wrm.rainforest Information on the threats to forests and their peoples, and about official solutions and NGO responses to them.

Green Movement

gpty.general Discussion about the Green Party and Green politics.

green.general Discussion and information about the international Green movement.

Health

en.health Wide-ranging discussion on the impact of the environment on health, and vice versa.

en.toxics Health and safety issues of toxic chemicals, their dangers and their control.

Media

gen.radio Discussion of radio and its applications in peace and environmental work.

gp.news A scanning of environmental news gleaned from major news wire services.

sc.natlnews The *Sierra Club National News Report*, a summary of news concerning the nation's environment published twice monthly by the Sierra Club.

Newsletters

en.consdigest The *Conservation Digest Newsletter*, published by the Munson Foundation. Concise information about issues and activities in natural resource conservation.

gp.natlnews Greenpeace's national newsletter, with information on its campaigns around the globe.

iucn.news News on species conservation, tropical forests, critical sites and protected areas, wetlands, population, environmental education, sustainable development, and computerized global conservation monitoring.

Populations

gen.nativeam Discussion of Native American issues.

gen.nativenet Discussion of issues related to indigenous peoples of the world.

gn.tribalsurvi Issues pertaining to the survival of indigenous peoples.

Seas and Waters

en.dolphinnet Marine mammal captivity issues. Commercial tuna netting procedures. Interspecies communication.

en.marine Information and discussion about the marine environment, including seas, creatures that inhabit them, and the impact of humankind on marine life.

en.water Domestic and international information and resources related to water pollution, coastal zone management, wetland protection, river conservation, and fish and wildlife.

Technology

elecenv Develop electronic search skills for the benefit of the environment. Includes listings and evaluations of online databases and a place to exchange tips and lessons learned.

gen.techeffect Effects of technology on health and the environment.

Toxics and Waste

en.recycle Information and discussion on recycling industrial and other wastes.

en.toxics.clea Toxic cleanup—sites and methods.

en.toxics.cont Toxic contamination—spills, accidents, and deliberate releases.

en.toxics.inci Hazardous waste incinerators; waste-to-energy plants.

en.toxics.land Land disposal of wastes.

gan.recycling Comprehensive information about recycling and what you can do to encourage it.

nuc.facilities Environmental effects of nuclear weapons production facilities, and actions organized against such facilities.

Wilderness and Wildlife

gan.wilderness Comprehensive information about wilderness issues and what you can do to help.

gn.animals News and discussion about all aspects of animal rights in farming and other areas.

All the people of a country have a direct interest in conservation.... Wildlife, water, forest, grasslands—all are part of man's essential environment; the conservation and effective use of one is impossible except as the others are also conserved.

Rachel Carson, 1946

GEnie

GEnie (General Electric Network for Information Exchange) is a commercial online service owned and operated by General Electric Company. GEnie has over 100,000 subscribers and offers mail, special interest groups, online shopping and travel services, games, live conferences, special interest "RoundTables," clubs, and software. GEnie is particularly innovative in its news service offerings, which are covered in detail in chapter 15.

How to Get Online

GEnie membership is open to anyone with a personal computer and a modem; you are not required to purchase any type of membership kit to sign on for the first time. All you need to do is follow these steps:

- Set your communications software to half duplex.
- Dial 800/638-8369
- When you are connected, type `HHH`.
- When you see the U#= prompt, type `XTX99515,GENIE` and then press ENTER or RETURN.

201

♦ Follow the online instructions.

If you would like to get information about GEnie services and pricing before you sign on for the first time, call 800/638-9636, and you will be sent an information kit. Once you have signed on, you have the opportunity to order a GEnie user manual.

Cost

GEnie charges a $4.95 monthly membership fee that allows unlimited use of certain services, including mail and some entertainment, reference and news services. For other services, you pay $6 per hour (Monday through Friday 6 p.m. to 8 a.m. local time, all day on weekends and holidays) or $18 per hour (8 a.m. to 6 p.m. weekdays). Certain services and databases are subject to additional charges.

Organization

Figure 11-1 shows the GEnie main menu. To go to any of the services listed there, simply type the menu item number at the prompt.

Getting Around

GEnie is menu driven but lets you use keywords to save time. At the top of each menu you find keywords and page numbers. To get to where you want to go, you either type a keyword at the prompt—for example, `mail` to get to the mail menu—or you type a page number with the word *move*, for example `Move 200`.

Environmental Resources

The following services on GEnie are of interest to environmental researchers:

FIGURE 11-1

```
                                    GENIE
No letters waiting.


        GEnie Announcements (FREE)

    1. Help out GEnie, take our free survey today................*SURVEY
    2. Celebrate SHERLOCK HOLMES' birthday tonight in................WRITERS
    3. Meet Jim Bensberg, the AMA's Washington Lobbyist   =o&o>......MOTO
    4. Walt Disney World's NEWEST project is making headlines in ->..*FLORIDA
    5. Get the Jump on Uncle Sam: Buy Tax Software NOW...............EXPRESS
    6. NEW Mac software library lists now =available= in............MAC
    7. In the know about UFOs in....................................PSI-NET
    8. If you're a new GEnie User, see us in the GEnie Users' RT,...GENIEUS
    9. If you play RSCARDS GAMES we have a Category for you in......MPGRT
   10. The snow's been falling, it's time to SKI in.................CALIFORNIA
   11. Check out what's happening in the Unix RT....................UNIX
   12. Everything you wanted to know about.........................ASTROLOGY
   13. Get to Know Lands' End: Order a Catalog Today................LANDSEND
   14. It's one of the funnest games around, join others playing....TRIVIA

Enter #, <H>elp, or <CR> to continue?

GEnie                         TOP                         Page    1
                      GE Information Services

    1.   GEnie*Basic Services              2.[*]GEnie Information
    3.[*]Billing and Setting Information   4.   Communications (GE Mail & Chat)
    5.   Computing Services                6.   Travel Services
    7.   Finance & Investing Services      8.   Online Shopping Services
    9.   News, Sports & Features          10.   Multi-Player Games
   11.   Career/Professional Services     12.   Business Services
   13.   Leisure Pursuits & Hobbies       14.   Education & Reference Services
   15.   Entertainment Services           16.   Symposiums on Global Issues
   17.   Leave GEnie (Logoff)

Enter #, <H>elp?

     1:42:31 PM
```

GEnie's main menu. To select an item, type its number at the prompt.

- The Medical RoundTable
- Electronic mail
- News wire services (see chapter 15)

The Medical RoundTable

GEnie's special interest groups are called RoundTables. GEnie RoundTables, like CompuServe's forums, offer public message areas, file libraries, real-time chatting, and an area where the SysOp can place announcements.

The Medical RoundTable, found in the Reference Library and hosted by Michael P. Weinstein, provides a forum for discussing medical and public health issues with physicians, health professionals, and other interested GEnie members. More than a dozen file libraries cover AIDS, basic science and research, public health, nutrition, and drug information. Discussion areas are arranged by category, including occupational and environmental health. This forum is an excellent place to discuss environmental health issues.

To get to the Medical RoundTable, first type `reference` or `Move 301` at the prompt to get to the Reference Library, then enter `MEDICAL` or `Move 745`. Figure 11-2 shows the Medical RoundTable menu. It does not cost anything to join this forum.

Electronic Mail

GEnie's electronic mail system allows you to send mail to any GEnie member. You can send files along with your messages, using the Xmodem file transfer protocol. Type `mail` or `Move 200` to send mail.

FIGURE 11-2
```
The Medical Roundtable
GEnie     MEDICAL     Page 745
    Medical RoundTable
    Library: ALL Libraries
 1. Medical Bulletin Board
 2. Medical Real-Time Conference
 3. Medical Software Libraries
 4. About the RoundTable
 5. RoundTable News          -900602
 6. Issues and News in Medicine 900527
Enter #, <P>revious, or <H>elp?
```

The Medical RoundTable main menu.

My country is the world. My countrymen are all mankind.

William Lloyd Garrison

CHAPTER 12

The WELL

The WELL (Whole Earth 'Lectronic Link) is a "conferencing" network run out of the San Francisco Bay Area offices of the *Whole Earth Catalog* and the *Whole Earth Review*. As a conferencing system, The WELL is composed of a series of discussions, or conferences, on a wide range of issues. A sense of community, freedom of expression, and open debate are all encouraged.

Unlike online services that offer a variety of information databases, The WELL's sole speciality is the art of online conversation. The WELL is very people-oriented. In addition to participating in conferences, you can chat "live" with other users online and download and upload files.

How to Get Online

Membership in The WELL is open to all; and you can access the service with your own communications software. To register, use your communications software to dial 415/332-6106. To have information sent to you before you try to connect, call 415/332-4335.

Cost

The WELL charges a monthly membership fee of $10 plus an online fee of $2 per hour (for any calls made anytime). If you dial the Sausalito, California number listed previously, you will also pay all applicable long-distance fees. You are allocated 500K of disk space free but charged $20 per month per megabyte for additional storage space.

As an alternative to dialing into The WELL long-distance, you can use a local phone number from the CompuServe Packet Network at an additional cost of $5 per hour inside the continental United States or $9 per hour in Alaska and Hawaii. To find out about less-expensive telephone charge alternatives, see "Saving on Phone Costs" in chapter 2 of this book.

Organization

The WELL is divided into fourteen major conference sections:

- Social Responsibility and Politics
- Media and Communications
- Business and Livelihood
- Body/Mind/Health
- Cultures
- Place
- Interactions
- Arts and Letters
- Recreation
- Entertainment
- Education and Planning
- Grateful Dead

- Computers
- The WELL Iiself

Each major conference is divided into more specific conferences, which are subdivided into "topics" and "responses" to that topic.

Getting Around

Since The WELL operates on the UNIX operating system and is command driven, you need to be specific about typing the right commands for navigation. Instead of seeing menu items, you see an OK: prompt. Entering commands at the prompt takes you to the various areas of interest. For example, by typing `?conf`, you can get a list of conferences (see Figure 12-1). By typing `?`, you can get online help. Typing `exit` at the OK: prompt takes you off the system.

Fortunately, an extensive user manual is available to download from The WELL (you need it). To download the manual into your computer, type the following at the OK: prompt:

```
!xm st /well/info/manual/fullmanual
```

When you sign onto the service for the first time, you are sent a WELL basic command card and a tutorial that helps you find your way around The WELL.

FIGURE 12-1

```
****    CONFERENCES      ****
Type:   cfinfo confname <cr> for a description of a
        conference. (example: cfinfo peace)

        Best of The WELL - vintage material      (g best)
        WELL "Screenzine" digest                 (g zine)
        Index listing of new topics in all conferences
                                              (g newtops)
```

Social Responsibility and Politics

Amnesty International	(g amnesty)	Non Profits	(g non)
Current Events	(g curr)	Peace	(g peace)
Firearms	(g firearms)	Politics	(g pol)
First Amendment	(g first)	Telecom Law	(g tcl)
Gulf War	(g gulf)	Veterans	(g vets)
Liberty	(g liberty)		

    Electronic Frontier Foundation   (g eff)
    Computers, Freedom & Privacy   (g cfp)
    Computer Professionals for Social Responsibility (g cpsr)

Media and Communications

Bioinfo	(g bioinfo)	Photography	(g pho)
Computer Journalism	(g cj)	Radio	(g rad)
Info Age	(g boing)	Technical Writers	(g tec)
Media	(g media)	Telecommunications	(g tele)
Microtimes	(g microx)	Usenet	(g usenet)
Muchomedia	(g mucho)	video	(g vid)
Netweaver	(g netweaver)	Virtual Reality	(g vr)
Networld	(g networld)	Whole Earth Review	(g we)
Packet Radio	(g packet)	Zines/Factsheet Five	(g f5)
Periodical/Newsletter	(g per)		

Business and Livelihood

Agriculture	(g agri)	Legal	(g legal)
Classifieds	(g cla)	One Person Business	(g one)
Consultants	(g consult)	The Future	(g fut)
Consumers	(g cons)	Translators	(g trans)
Entrepreneurs	(g entre)	Work	(g work)
Investments	(g invest)		

Body - Mind - Health

Aging	(g gray)	Jewish	(g jew)
AIDS	(g aids)	Men on The WELL*	(g mow)
Buddhist	(g wonderland)	Mind	(g mind)
Christian	(g cross)	Philosophy	(g phi)
Dreams	(g dream)	Psychology	(g psy)
Emotional Health**	(g private)	Recovery**	(g recovery)
Erotica	(g eros)	Sexuality	(g sex)
Fringes of Reason	(g fringes)	Spirituality	(g spirit)
Health	(g heal)	Women on The WELL#	(g wow)
Holistic	(g holi)		

*   Private conference - mail flash for entry
**   Private conference - mail wooly for entry
***   Private conference - mail dhawk for entry
\#   Private conference - mail reva   for entry

Cultures

Archives	(g arc)	Spanish	(g spanish)
Buddhist	(g wonderland)	Pacific Rim	(g pacrim)
German	(g german)	Tibet	(g tibet)
Italian	(g ital)	Travel	(g tra)
Jewish	(g jew)	History	(g hist)

Place

Berkeley	(g berk)	Northwest	(g nw)
East Coast	(g east)	Pacific Rim	(g pacrim)
Environment	(g env)	Peninsula	(g pen)
Geography	(g geo)	San Francisco	(g sanfran)
Hawaii	(g aloha)	Southern USA	(g south)
North Bay	(g north)	Tibet	(g tibet)

Interactions

Couples	(g couples)	News	(g news)
Disability	(g disability)	Nightowls##	(g owl)
Gay	(g gay)	Parenting	(g par)
Gay (private)#	(g gaypriv)	Scams	(g scam)
Interview	(g inter)	Singles	(g singles)
Kids 91	(g kids)	True Confessions	(g tru)
Miscellaneous	(g misc)	Unclear	(g unclear)
Weird	(g weird)		

\#   Private conference - mail   hudu   for entry
\#\#   Open from midnight to 6 am

Arts and Letters

Art Com			
Electronic Net	(g acen)	Photography	(g pho)
Art and Graphics	(g gra)	Poetry	(g poetry)
Books	(g books)	Radio	(g rad)
Comics	(g comics)	Science Fiction	(g sf)
Design	(g design)	Bay Area Siggraph	(g siggraph)
MIDI	(g midi)	Theater	(g theater)
Movies	(g movies)	WELL Writer's Workshop*	(g www)
Muchomedia	(g mucho)	Words	(g words)
NAPLPS	(g naplps)	Writers	(g wri)
On Stage	(g onstage)	Zines/Factsheet Five	(g fs)

\#   Private conference - mail   sonia   for entry

Recreation

Bicycles	(g bike)	Gardening	(g gard)
Boating	(g boat)	Motorcycling	(g ride)
Cooking	(g cook)	Motoring	(g car)
Flying	(g flying)	Pets	(g pets)
Games	(g games)	Sports	(g sports)

Entertainment

Audio-videophilia	(g aud)	Movies	(g movies)
Bay Area Tonight#	(g bat)	Music	(g music)
CD's	(g cd)	Restaurant	(g rest)
Comics	(g comic)	Star Trek	(g trek)
Fun	(g fun)	Television	(g tv)
Jokes	(g jokes)		

\# Updated daily

Education and Planning

Apple Library User's Group	(g alug)	Homeowners	(g home)
Brainstorming	(g brain)	Indexing	(g index)
Design	(g design)	Network Integrations	(g origin)
Education	(g ed)	Transportation	(g transport)
Energy	(g energy91)	Whole Earth Review	(g we)
Science	(g science)		

Grateful Dead

Grateful Dead	(g gd)	Deadplan*	(g dp)
Deadlit	(g deadlit)	Feedback	(g feedback)
GD Hour	(g gdh)	Tapes	(g tapes)
Tickets	(g tix)	Tours	(g tours)

* Private Conference - mail tnf for entry

Computers

AI/Forth/Realtime	(g realtime)	NAPLPS	(g naplps)
Amiga	(g amiga)	NeXt	(g next)
Apple	(g apple)	OS/2	(g os2)
Art and Graphics	(g gra)	Printers	(g print)
Computer Books	(g cbook)	Programmer's Net	(g net)
Desktop Publishing	(g desk)	Bay Area Siggraph	(g siggraph)

```
Hacking           (g hack)       Software Design        (g sdc)
Hypercard         (g hype)       Software/Programming   (g software)
IBM PC            (g ibm)        Software Support       (g ssc)
Lans              (g lan)        Unix                   (g unix)
Laptop            (g lap)        Virtual Reality        (g vr)
Macintosh         (g mac)        Windows                (g windows)
Mactech           (g mactech)    Word Processing        (g word)
MIDI              (g midi)

                                 The WELL Itself
                                 ---------------
Deeper            (g deeper)     Hosts                  (g host)
Entry             (g ent)        Policy                 (g policy)
General           (g gentech)    System News            (g sysnews)
Help              (g help)       Test                   (g test)
```

A list of the conferences displayed when you type ?conf *at the* OK: *prompt.*

Environmental Resources

The WELL offers the following resources for environmentalists:

- The Environment conference, located in the Place section
- The Science conference, located in the Education and Planning section
- The BioInfo conference, located in the Media and Communications section
- The Usenet conference, where you can access Usenet newsgroups, located in the Media and Communications section
- Electronic mail, which can connect you to the Internet and BITNET

Conferencing on The WELL

Conferencing—the open exchange of opinions on a wide variety of topics—is the heart of The WELL. The WELL is dynamic and driven by the interests of its members. Conferences on The WELL of use to those interested in environmental issues include Environment, BioInfo, and Science. This section takes an in-depth look at the Environment conference.

The Environment Conference

The Environment conference is located in the Place section of The WELL.

Entering a conference is as simple as typing g and the conference keyword. While in a conference, you can get a list of the topics, read any or all of them, respond to a topic, or create a new one. Figure 12-2 takes a look at the Environment conference.

FIGURE 12-2

```
Ok (? for help): g env
  "Inspect every piece of pseudoscience and you will find a
security blanket, a thumb to suck, a skirt to hold. What
have we to offer in exchange? Uncertainty! Insecurity!"
           "Isaac Asimov
  ***DESIGN PHASE: WELL's new Environmental Conference ***
12 newresponse topics and 58 brandnew topics
First topic 1, last 109
You have mail.
Ok (? for help): b
Topic - Number of responses - Header
  1    1 Designing the Environmental Conference: an
Invitation--
  3   90 Let's Introduce Ourselves: who, what, why, when,
where, and so on.
  5   10 The Chang Tang
```

```
 6    4 Critter Cover: Planting for Animals
 7   24 Gashed by Its Own Anchor: Oil Spill off the California Coast
 8  108 Animal Rights Activists as Environmentalists
 9   43 What Does "Natural" Mean?
10   24 Environmental groups
11   32 The Environment in the Media
12  155 Forest Issues - California and the Pacific Northwest
13    3 Where to recycle/dispose of used motor oil and other toxics?
14   15 Bovine Growth Hormone
15   21 Chief Seattle's Message—1854
16   26 Teak Harvesting & Management—What do you know?
17   70 How does one balance global environmental concerns against
individual health?
19    0 CALIFORNIA'S ENDANGERED SPECIES
22   26 Kitchen Table
23   34 Acid Rain—two contradictory reports in the news 2/20/90
24    6 Coming Events
25    0 Restoration: special issue of Whole Earth Review (in the 'we'
conference)
26    2 The Tragedy of the *Unmanaged* Commons (Garrett Hardin)
27    7 The Clean Air Act Amendments
28    2 Are redwood burls an endangered species?
29   14 Carrying Capacity - A legal concept?
30  296 Spiritual Environmentalism
31   23 Jeremy Rifkin—Environmentalist or Quack
32   60 RECYCLING:  for love or money?
34    2 WARNING: Grassroots Activism at Work! In Los Angeles...
35   35 Personal favorites: books, newspapers, and journals
36    3 Forests Forever Initiative Needs your Help!
37   14 New Paradigm Needed for Environmental Action!
38    4 Globe 90 Environment/Business Conf - Vancouver March 19-23
39   43 Organic Gardening/Sustainable Agriculture: Respect Your Mother!
40  149 Problems with Nuclear Power.
41   81 Carcinogens
42   28 Dangerous Sunshine
```

43 130 Are 'non-native' plant species harmful to an ecosystem?
44   26 Indoor Environmentalism
45   16 What got (will get) you to act environmentally-responsible?
46   90 The Malathion Controversy
47   1 Recyling theory & practice: pros, cons, and caveats.
48   27 Media on Environment
50   28 Prescribed burning
51   29 Case Study #5   A New General Store—-Bulk items only!
52   2 Problems with Conservation
53   32 fireplaces...beauty and the beast
54   0 Ecology Books Desperately Needed in POLAND
55   6 Paper Recycling Technology Needed in POLAND
56   72 One Closed Living System: Biosphere II
57   13 Why do people live in the country?
58 163 Environmental Protest
59   25 Cultural and environmental survival
60   6 Usenet newsgroups relating to environmental concerns
61   45 Economy and Environment
62   74 Mass Transit
63   1 Actions To Help Preserve Wild and Scenic Rivers
64   10 Interior Secretary Lujan Attacks Endangered Species
65   42 Toward an Environmental Energy Policy
66   13 US Fends Off Global Climate Action
67   4 Offshore drilling gives me that sinking feeling
68   73 FOREST ISSUES - 1990
69   7 Options - new conference and magazine
70   9 American Chestnut Foundation
71   16 Population Control - Necessary? Possible?
72 125 Defining Sustainability
73 268 EARTH FIRSTIANS DETERMINE TO SAVE THE PLANET: ARE BOMBED
74   59 Can of Worms—Asbestos in Remodeling
75   8 Environmental "restoration": next year's magic word?
76   19 Eastern Europe: Ecological Crisis Meets Political Crisis
77 113 Earth First Bombing — News and New Developments
78   7 Redwood Summer

```
79   19 Siting a landfill??
80    5 Contacts in E. Europe?
81   59 Global Climate Change — in our time?
82    2 REDWOOD SUMMER DEMONSTRATION AT FORT BRAGG & MORE !
84    1 Birds and Bugs, a necessary unity
85    9 Oral Contraceptives: What a Waste!
86   10 Making Trade-Offs/Living Lightly/Urban Living
87   16 The Journey Home
88    0 Asbestos in drinking water—Cause cancer??
89    2 Suggestions on "environmentally responsible" housecleaning
services?
90   28 PATRICIA POORE INTERVIEW
91   27 PATRICIA POORE DISCUSSION
92   32 Environmental Art & Artists
93   14 Environmental addresses/phone nos.: topic for the actively
concerned.
94    8 Wind Energy/Global Warming News
95    8 Endangered Tibet
96    2 Information Wanted on CERES and The Valdez Principles
97    8 Save our Seas-South Pacific
98   36 Re-opening of LOVE Canal
99   50 "nuclear waste" that is "not of regulatory concern"?
100   7 The Bullshit Starts: a record-keeping topic for the CA
environmental issues.
101   9 A Guessing Game?
102  12 ECONET
103   1 Poison gas touches a "nerve" in the South Pacific
104  14 OCTOBER - A HARVEST OF DISSENT
105  56 Earth Action Reponds - The Case for Green Direct Action !
106  12 Foxproofing a cat box
107   0 BRC ALERT - NRC Public Hearings.
108   2 Sharing our environmental ethos
109   7 Organic Pest Control
Ok (? for help): s1
Topic   1: Designing the Environmental Conference: an Invitation—
```

By: Penny Post (macpost) on Wed, Jan 31, '90
    1 responses so far

In response to a terrific resurgence in popular demand, we're starting up The WELL's Environmental Conference again. This time, though, we're looking for your participation in designing the conference...we'd like to learn what's on *your* mind right now, what environmental concerns and bugaboos you want to read and write about, your special areas of expertise.
So we'd like you to open topics as you feel inspired, and after a few weeks, we'll take a look at what's here, organize it, and put it into conference form. By the way, if you've never opened a topic before, it's not at all hard--just take a look at the new WELL Manual, pp. 57-58, "Creating a New Topic".

    1 response total.

The Environment conference. User input is boldfaced. Typing `g env` *selects the Environment conference. Entering* `b` *for "browse" enables you to examine a list of the topics in the conference. Notice that each topic includes a series of responses, or comments from members of The WELL. Typing* `s` *for "see" followed by the number of a topic in which you are interested displays an introduction to that topic. After you read the topic description, you can either respond to the topic (that is, add your two cents to the topic under discussion), or read the responses of other members of The WELL.*

The Science Conference

The Science conference is located in the Education and Planning section of The WELL. To get to this conference, type `g science` at the OK: prompt.

The BioInfo Conference

The BioInfo conference is located in the Media and Communications section of The WELL. To get to this conference, type `g bioinfo` at the OK: prompt.

Usenet

Chapter 5 has already familiarized you with Usenet newgroups. If you are unable to access Usenet newsgroups from a public UNIX BBS, EcoNet, or an academic institution, you can access them through The WELL. Enter the Usenet conference by typing `g usenet` at the OK: prompt.

Electronic Mail

In addition to sending e-mail to other members of The WELL, you can use The WELL's mail system to send mail to users of Internet and BITNET.

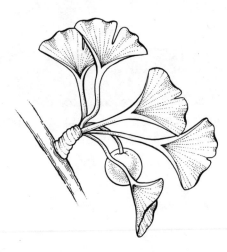

PART V

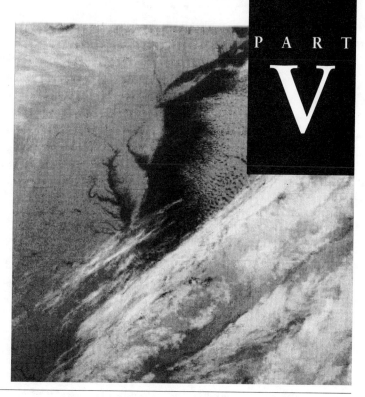

Libraries That Never Close

Before the introduction of personal computers, creating bibliographies and finding reference material for environmental research meant spending many nights in your local library thumbing through volumes of printed indexes and card catalogs, painstakingly writing down each citation, then finding the document and photocopying it. Over the last decade, online bibliographic database companies have made researching easier, and most libraries have installed computer terminals that allow a quick search of card catalogs.

Online database retrieval companies have now taken the next step forward in their evolution by giving the researcher—at home or in the office—access to millions of articles at the touch of a keyboard nearly 24 hours a day, seven days a week. Low-cost bibliographical and database retrieval services provide access to many of the most popular databases covering science and technology, the arts, the humanities, and business, as well as specialized sets of scientific collections. For a low fee, anyone with a computer, a modem, and communications software can conduct extensive bibliographical research, from the convenience of home or office.

Chapter 13 describes three commercially available online database services that together give you access to nearly every environmental and scientific publication in existence:

- Knowledge Index (KI)
- BRS/After Dark
- ORBIT

In addition, the chapter describes specialized databases that further broaden your access to bibliographic services and technical information.

Chapter 14 shows you how the recent advances in CD-ROM technology can offer you a relatively inexpensive and efficient way to keep important environmental resources at your fingertips.

Chapter 15 looks at online news services that allow you to cut through information clutter by accessing "hot-off-the-wire" environmental news that interests you—at any time of the day or night—before it hits your local newspaper or television news.

Our ideals, laws and customs should be based on the proposition that each generation in turn becomes the custodian rather than the absolute owner of our resources—and each generation has the obligation to pass this inheritance on in the future.

Alden Whitman, 1971

CHAPTER 13

Online Research Databases

Knowledge Index

Dialog Information Services, which offers Knowledge Index, is the largest online bibliographic company in the United States, with over 200 million items available for searching in 370 databases. Dialog was created in 1963 out of a research and development program at Lockheed Missiles and Space Company. In 1972, Dialog became a commercial search service and a wholly owned subsidiary of Lockheed Corporation. Knight-Ridder, Inc., purchased the company in 1988.

In 1981, Dialog launched Knowledge Index to meet the needs of personal computer users interested in science and education. Offering evening and weekend hours to keep online costs to a minimum, Knowledge Index allows you to search its database using keywords, phrases, and Boolean connectors to create a bibliography.

How to Get Online

To get information about subscribing to Knowledge Index, call Dialog Information Services at 800/334-2564. You can use any communications software to dial a local SprintNet or Tymnet

Revising the Toxic Release Inventory

David Abercrombie, a chemical engineer, feels almost offended by the lack of ranking by relative toxicity of the 300 chemicals on the Environmental Protection Agency's Toxic Release Inventory (TRI) list. In his view, EPA's decision to treat the chemicals as toxic equals, pound for pound, is not only the product of lazy thinking, but inaccurate and potentially dangerous. For example, one chemical on the list, Dibenzofuran, is over ten million times more toxic per pound than another chemical on the list, Di(2-ethylhexyl)phthalate.

Under EPA regulation, any place of business that uses more than 10,000 pounds of any chemical on the TRI list must submit a separate annual report for each chemical on the list. EPA compiles this data into a database, and stores it on the Toxnet computer system, managed by the National Library of Medicine. The TRI database does not include any toxicity data; it only lists the weight of each chemical that is used or released.

Abercrombie feels this list of sites and chemicals is not very useful. He decided to develop a model ranking system of sites, based on total pounds of release of each chemical multiplied by an individualized toxicity factor for each chemical.

Abercrombie started with Toxnet, the name given to EPA's list of sites and emissions on the Medlars Computer System. "This information is also available on paper, but it's too difficult to use that way," he says. "You can't find what you're looking for."

He downloaded the whole database to his hard drive, and cleaned up its records a bit. He consolidated them so there was one per facility, listing all the chemicals that facility uses.

Next, he looked for a reasonable way to rank the TRI chemicals, based purely on human health. For that, Abercrombie chose the IRIS list of toxicity numbers, also on the Medlars system. Only 51 chemicals on the TRI list are listed in IRIS. He supplemented this list with California State guidance documents used in the "Air Toxics Hotspots" program. Abercrombie could not find reliable numbers for the rest of the 300 chemicals listed in the Toxic Release Inventory.

Abercrombie then imported everything—the data on the sites, the chemical names, and the relative toxicity numbers—into Paradox, a database program. He added up the toxicity factors of all the chemicals each site releases, and was able to come up with his own ranked list. The resulting database can be searched by zip codes, chemicals, or name of a given facility. To his surprise, his database found the highest ranked site in the United States to be a manufacturer of processed foods in Chicago.

He then presented his database at a conference of environmental consultants. Abercrombie hopes to get grants from state agencies and the EPA to continue his research, and eventually to bring about a change in the state law governing reporting of toxic emissions by industry, Law AB2588, to also require a risk assessment. So far, no one has gathered the necessary data to make a risk assessment meaningful, he says. "Once someone comes up with the numbers, the State will feel compelled to change the law."

Abercrombie can be reached on the UseNet at abe@lyra.infoserv.com.

—Wendy Monroe

node or Dialog's own Dialnet node to connect to Dialog's Hitachi mainframes in Palo Alto, California. Alternatively, Dialog offers its own software for IBM and compatibles called DialogLINK. A series of scripts for logging on to Knowledge Index is available from Desktop Information (32 W. Anapoamu, Ste. 200, Santa Barbara, CA 93101, 805/963-4095) for users of MicroPhone, a Macintosh communications program from Software Ventures.

Cost

New subscribers pay a start-up fee of $35. The fee includes two hours of free time, a user manual containing a list of all Knowledge Index databases and their labels, sample searches, and other information, along with detailed explanations of the various search commands and shortcuts.

Online fees are $24 per hour, at off-peak times; the fee includes all connecting network phone charges. No fee is charged per citation.

Instead of visiting a local library to make copies of articles, you can order the full text from Dialog for a fee of $7.50 per article plus 35 cents per photocopied page. Even better, you can capture the text into your computer to print later.

Online Availability

Monday through Thursday, 6 p.m. to 5 a.m.; weekends (6 p.m. Friday to 5 a.m. Monday). Available worldwide.

Knowledge Index Databases

The following are just a few of Knowledge Index's databases that might be of use to environmental researchers. Brief descriptions of the databases' contents, and data on how frequently each is updated, are included.

Academic American Encyclopedia

The Academic American Encyclopedia is a complete encyclopedia including over 34,000 articles written by 2,300 contributors who are experts in their fields. The articles average 300 words in length. Cross-references provide quick access to related subjects, and bibliographies are provided for longer articles. Fact boxes give key information on subjects such as presidents, states, and countries.

> Coverage: Current edition
> Updated: Quarterly

Academic Index

The Academic Index covers more than 400 scholarly and general interest publications, including the most commonly held titles in over 120 college and university libraries. Subjects covered include art, anthropology, economics, education, ethnic studies, government, history, literature, political science, general science, psychology, religion, sociology, and leisure.

> Coverage: 1976 to present
> Updated: Monthly

Agricola

The database of the National Agricultural Library provides worldwide coverage of journals and monographs on agriculture and related subjects such as animal studies, botany, chemistry, entomology, fertilizers, forestry, hydroponics, and soils.

> Coverage: 1970 to present
> Updated: Monthly

Agribusiness USA

The Agribusiness USA database provides controlled-vocabulary indexing and informative abstracts from approximately 300 industry-related trade journals and government publications.

> Coverage: 1985 to present
> Updated: Every two weeks

Agrochemicals Handbook

The Agrochemicals Handbook provides information on the active components of agrochemical products used worldwide. For each substance, The Agrochemicals Handbook gives the following information: chemical name, including synonyms and trade names; CAS Registry Number; molecular formula; molecular weight; manufacturer's name; chemical and physical properties; toxicity; mode of action; activity; and health and safety concerns.

> Coverage: Current
> Updated: Semianually

Books in Print

Books in Print is the major source of information on books currently in print in the United States. The database provides a record of forthcoming books, books in print, and books out of print. Scientific, technical, medical, scholarly, and popular works, as well as children's books, are included in the file.

> Coverage: Currently in-print books
> Updated: Monthly

CAB Abstracts

CAB Abstracts is a comprehensive file of agricultural and biological information containing all the records from the 26 main abstract journals published by the Commonwealth Agricultural Bureau. Over 8,500 journals in 37 languages are scanned for inclusion, as are books, reports, theses, conference proceedings, patents, annual reports, and guides.

> Coverage: 1972 to present
> Updated: Monthly

CompendexPlus

Compiled from over 4,500 sources, CompendexPlus contains information on soil contamination, groundwater, pollution control equipment, and water treatment. It is considered the most comprehensive source available for engineering.

> Coverage: Current
> Updated: Monthly

Consumer Drug Information Fulltext

Consumer Drug Information Fulltext (CDIF) includes the complete text of the *Consumer Drug Digest,* published by the American Society of Hospital Pharmacists. CDIF contains in-depth descriptions of more than 260 drugs, including over 80 percent of all prescription drugs and a number of important nonprescription drugs. Information includes how the drugs work; possible side effects and how to manage them; precautions; instructions on dosage and use, including the foods and activities that are

Thwarting International Recycling Fraud

Mark Walton (not his real name), an environmental engineer, got an urgent call from a friend who worked for a well-known environmental activist group. She had heard of a suspicious-sounding recycling scheme, and she wanted to know whether Walton could find out whether it was legitimate or not.

A United States chemical recycling firm had applied for a permit to begin exporting an annual 5000 tons of electric arc furnace dust to a recycling facility in the Atacama desert in Chile.

Electric arc furnace dust, which collects in filters in steel mills, contains zinc, copper and heavy metals such as lead and cadmium. The Environmental Protection Agency designates it as hazardous waste, and regulates its handling and storage from manufacture to disposal. Many American steel mills have researched methods to recycle this dust and have concluded that it is not economically sensible to recycle it in the United States.

The brochure from the Chilean recycling facility said the plant is located on a former military reservation, and, as part of the process, recycles sludge containing the dust in solar evaporation ponds. Since it hasn't yet rained in the Atacama desert in recorded history, Walton thought this part of the plan sounded reasonable.

The Chilean recycling company had supplied detailed business plans to the American steel mill, including a description of the plant and its method of operations. Walton tried to figure out the feasibility of the plans: he started with the figures given for the cost of operating the plant, divided this by the number of pounds the plant could process, and came up with a recycling cost of $1.50 per pound, including transportation to Chile by barge.

> He next researched the Chilean metals industry by searching Knowledge Index. He read about the feasibility of recycling arc furnace dust on the Chemical Business Newsbase.
>
> Walton soon realized that Chile is a major metals exporter, and it costs about 40 cents per pound in that country to dig zinc ore and process it into metal. It became clear that reclaiming zinc and copper from the dust for $1.50 per pound made no economic sense. He concluded the Chileans were not intending to recycle the dust at all, but were planning to dump it in violation of EPA statutes and collect the disposal fee from the United States.
>
> Walton told his activist friend of his conclusion, and her group sent a series of faxes to activists in Chile alerting them of the proposed toxic dumping scheme. Later, a series of hearings were held in Chile to prevent the arc furnace dust "recyclers" from soliciting any further business.
>
> —Wendy Monroe

permitted during medication and what to do if a dose is missed; and advice on storing drugs.

> Coverage: Current
> Updated: Quarterly

Consumer Reports

Consumer Reports contains the complete text of the 11 regular monthly issues of the print *Consumer Reports* and the 12 monthly issues each of the *Consumer Reports Travel Letter* and the *Consumer Reports Heath Letter*.

> Coverage: 1982 to present; *Consumer Reports Health Letter*, September 1989 to present.
> Updated: Monthly

Current Biotechnology Abstracts

The Current Biotechnology Abstracts is an online version of the print publication of the same name; both are produced by the Royal Society of Chemistry, and the online database contains all the material published in the hard-copy version since its start-up in April 1983. Subjects include genetic manipulation, monoclonal antibodies, immobilized cells and enzymes, single-cell proteins, and fermentation technology, and the applications of those technologies in industries such as pharmaceuticals, fuel, agricultural chemicals, and food.

> Coverage: 1983 to present
> Updated: Monthly

Dissertation Abstracts Online

Dissertation Abstracts Online is a definitive guide to virtually every American dissertation accepted at an accredited institution since 1861, when academic doctoral degrees were first granted in the United States. In addition, Masters Abstracts from Spring 1988 to the present are included.

> Coverage: 1861 to present
> Updated: Monthly

Economic Literature Index

The Economic Literature Index covers journal articles and book reviews from 260 economics journals and approximately 200 monographs per year. Since June 1984, abstracts from selected journals have been added to approximately 25 percent of the records in the file. The descriptive abstracts are approximately 100 words in length and are written by the author or editor of the journal article; all are in English. The database corresponds to the index section of the quarterly *Journal of Economic Literature* and the annual *Index of Economic Articles*.

> Coverage: 1969 to present
> Updated: Quarterly

Engineering Literature Index

The Engineering Literature Index database is the machine-readable version of the *Engineering Index*, which provides abstracted information from the world's significant engineering and technological literature. The Engineering Literature Index database provides worldwide coverage of approximately 4500 journals and selected government reports and books. Subjects covered include civil, energy, environmental, geological, and biological engineering; electrical, electronics, and control engineering; chemical, mining, metals, and fuel engineering; mechanical, automotive, nuclear, and aerospace engineering; and computers, robotics, and industrial robots.

> Coverage: 1970 to present
> Updated: Monthly

Food Science and Technology Abstracts

Food Science and Technology Abstracts (FSTA) provides access to research and new developments in food science and technology and allied disciplines such as agriculture, chemistry, biochemistry, and physics. Other related disciplines such as engineering and home economics are included when relevant to food science. Also included in the file is Vitis, a subfile on viticulture and enology. Information in this subfile covers grapes and grapevine science and technology. FSTA indexes over 1200 journals from over 50 countries, patents from 20 countries, and books in any language.

> Coverage: 1969 to present
> Updated: Monthly

GPO Publications Reference File

The GPO (Government Printing Office) Publications Reference File indexes public documents currently for sale by the Superintendent of Documents, U.S. Government Printing Office, as well as forthcoming and recently out-of-print publications.

> Coverage: 1971 to present
> Updated: Biweekly

Heilbron

Heilbron, a chemical properties database, includes the complete text of two major chemical dictionaries from Chapman and Hall, Ltd.: the *Dictionary of Organic Compounds* (fifth edition), and the *Dictionary of Organometallic Compounds*. Also included are other source books, including: *Carbohydrates,*

Amino Acids and Peptides, The Dictionary of Antibiotics and Related Compounds, and *The Dictionary of Organophosphorus Compounds.* Heilbron helps identify chemical substances based on their physical and chemical properties, compound variants, derivative names, synonyms, CAS Registry Numbers, molecular formulas and molecular weight, source statements, use and importance data, melting point, freezing point, boiling point, solubility, relative density, optical rotation, and dissociation constants. Chemical images can be displayed when you search with DialogLINK software.

> Coverage: Current
> Updated: Semiannually

Legal Resource Index

The Legal Resource Index covers over 750 key law journals and six law newspapers, plus legal monographs.

> Coverage: 1980 to present
> Updated: Monthly

Life Sciences Collection

Life Sciences Collection contains abstracts of information in the fields of animal behavior, biochemistry, ecology, endocrinology, entomology, genetics, immunology, microbiology, oncology, neuroscience, toxicology, virology and related fields. The collection is an online version of the 17 volumes of abstracts by the same name.

> Coverage: 1978 to present
> Updated: Monthly

Magazine Index

Magazine Index covers more than 435 popular magazines, providing extensive coverage of current affairs, the performing arts, business, sports, recreation and travel, consumer product evaluations, science and technology, leisure-time activities, and other areas.

> Coverage: 1959 to March 1970, 1973 to present
> Updated: Weekly

Marquis Who's Who

Marquis Who's Who contains detailed biographies on over 77,000 individuals. Top professionals in business, sports, government, the arts, entertainment, science, and technology are included. Data provided includes career history, education, creative works, publications, family background, current address, political activities and affiliation, religion, and special achievements.

> Coverage: Current
> Updated: Quarterly

MEDLINE

MEDLINE (MEDLARS onLINE), produced by the U.S. National Library of Medicine, is one of the major sources on biomedical literature. MEDLINE is the equivalent of three printed indexes: *Index Medicus, Index to Dental Literature,* and *International Nursing Index.* MEDLINE covers virtually every subject in the broad field of biomedicine, indexing articles from over 3000 journals published in the United States and 70 other countries.

> Coverage: 1966 to present
> Updated: Twice a month

National Newspaper Index

The National Newspaper Index provides front-to-back indexing of the *Christian Science Monitor,* the *New York Times,* and the *Wall Street Journal.* All articles, news reports, editorials, letters to the editor, obituaries, product evaluations, biographical pieces, poetry, recipes, columns, cartoons and illustrations, and reviews are included. In addition, the National Newspaper Index covers national and international news stories written by the staff writers of the *Washington Post* and the *Los Angeles Times.*

> Coverage: 1979 to present (1982 to present for the *Los Angeles Times* and *The Washington Post*)
> Updated: Monthly

Newsearch

Newsearch is a daily index of more than 2000 news stories, articles, and book reviews from over 1700 important newspapers, magazines, and periodicals. In addition Newsearch includes the Area Business Databank, which indexes and abstracts from over 100 local and regional business publications, and the complete text of PR Newswire.

> Coverage: Current month only
> Updated: Daily

Pollution Abstracts

Pollution Abstracts is a leading source of references to literature on pollution, its sources, and its control. The following subjects are covered: air pollution, environmental quality, noise pollution, pesticides, radiation, solid waste, and water pollution.

> Coverage: 1970 to present
> Updated: Bimonthly

BRS/After Dark

BRS/After Dark is a low-cost service that offers more than 130 databases covering a wide set of disciplines, including biomedicine, education, business, life science, arts and humanities, and current events. New databases are added frequently. Like Knowledge Index, BRS/After Dark is designed for individual and academic users who wish to contain costs by accessing the database during evening and weekend hours.

BRS/After Dark provides many of the same databases as Knowledge Index. Depending on the database searched, your costs on After Dark are more or less expensive than Knowledge Index, since database rates and display charges vary on After Dark, while Knowledge Index charges a flat rate and no display charges.

After Dark is menu driven, using many of the same common sense Boolean search commands as Knowledge Index, and allows you to locate quickly a summary of any database and its cost before use. You display the references in a short form that gives you just enough information (title, author, publication, and date) to find the document quickly in your library. You can read the entire text of other databases. For an additional fee, you can have the complete search mailed to you within 24 hours.

BRS also offers BRS Colleague and Search Service, for medical and health care professionals conducting research. The service offers daily updates of medical news stories, electronic mail, interactive bulletin boards, and access to over 100 other databases in business, science, education, and social sciences. Rates for the Colleague and Search Service are more expensive than After Dark's, ranging from $8 to over $100 per hour evenings and weekends. Daytime searching is available for higher costs.

How to Get Online

For information about subscribing to BRS/After Dark or any other BRS database service, call Information Technologies, a Division of Maxwell Online, Inc., at 800/955-0906. You can access After Dark using your own communications software and the SprintNet, Tymnet, or DataPac communications network.

Cost

A one-time subscription fee of $75 includes a user manual and a complete set of database information sheets. Online charges vary depending on which database you use, from a low of $8 per hour to high of $48 per hour (with a $12 per month minimum). Connect time and communication fees are included. Sometimes you pay additional charges per citation display—up to 45 cents—with the average cost around 10 cents per citation.

Online Availability

Monday through Friday, 6 p.m. local time to 4 a.m. Eastern time; Saturday, 6 a.m. to 2 a.m. Eastern Time; Sunday, 9 a.m. to 4 a.m. Eastern time.

BRS/After Dark Databases

BRS/After Dark provides many of the same databases that are available on Knowledge Index. The following databases of interest to environmentalists, however, are exclusive to After Dark:

Cambridge Scientific Abstracts Life Sciences
This superfile contains access to the following abstract services: the Life Sciences Collection, Aquatic Sciences and Fisheries Abstracts, Oceanic Abstracts, and Pollution Abstracts.

> Coverage: 1981 to present
> Updated: Monthly

Comprehensive Core Medical Library: Science
Online access to the full text of *Science* magazine.

> Coverage: April 1986 to present
> Updated: Weekly

Current Awareness in Biological Sciences
This bibliographic database covers all aspects of the biological sciences, including biochemistry, genetics, ecology, plant science, toxicology, and developmental biology.

> Coverage: 1983 to present
> Updated: Monthly

Current Contents: Agriculture, Biology, and Environmental Sciences
An online version of the printed reference by the same name. Features author abstracts and keywords.

> Coverage: Most current 12 months
> Updated: Weekly

Current Contents: Life Sciences
Online version of the printed edition.

> Coverage: Most current 12 months
> Updated: Weekly

Current Contents: Physical, Chemical, and Earth Sciences.
Online version of the printed edition.

> Coverage: Most current 12 months
> Updated: Weekly

Federal Register Abstracts
Very useful for keeping up with constantly changing federal environmental regulations.

> Coverage: 1986 to present
> Updates: Weekly

National Environmental Data Referral Service
This database, produced by the U.S. Department of Commerce, is an index of published and unpublished environmental data from public and private sources worldwide.

> Coverage: 1983 to present
> Updated: Weekly

Orbit

A division of Maxwell Online, the same company that owns BRS, Orbit includes more than 100 databases in chemistry, earth sciences, energy, engineering, environmental, science, and material science. It offers many databases not available elsewhere.

Orbit is command driven, and full Boolean searching is fast and easy. Search functions enable you to discover information such as who are the most prolific authors on a given subject. You can also use the same search terms on more than one database. Orbit also lets you retrieve only the data you needed from each record, and allows you to browse a set of records to select those that are of interest.

How to Get Online

For information about subscribing to Orbit, call Orbit Search Service at 800/45-ORBIT.

Cost

You pay an annual subscription fee of $40 per user. Online charges depend on the database used, ranging from a low of $10 per hour to a high of almost $200 per hour, with a $15 per month minimum. Connect time and communications fees using SprintNet or Tymnet cost extra. Orbit charges an additional fee per citation display, starting at 10 cents per citation.

Online Availability

Monday through Friday, all day except 9:45 to 10:15 p.m.; Saturday, all day until 9:45 p.m.; Sunday, 10 p.m. to midnight.

Orbit Databases

The following is a selection of databases available on Orbit that might be of interest to scientists and environmentalists.

American Men and Women of Science

An active register of U.S. and Canadian professionals in the physical and biological sciences. Public-health scientists, engineers, mathematicians, statisticians, and computer scientists are also included. A total of 160 major disciplines and 800 subdisciplines are classified. Biographical citations for 125,000 scientists include what the individual has accomplished, where and when he or she accomplished it, current affiliations, and contact information.

> Coverage: Current
> Updated: Every three years

Analytical Abstracts

The most comprehensive abstract source available on analytical chemistry. Items cover general, inorganic, organic, biochemical, pharmaceutical, food, agricultural, and environmental aspects of analytical chemistry, including computer and instrumental applications in analysis, and are taken from nearly 2000 international primary journal sources. Journal articles represent 98 percent of the file; other sources include books, conference papers and proceedings, standards, and technical reports.

> Coverage: 1980 to present
> Updated: Monthly

Aqualine

Covers the world's literature on water and wastewater technology and environmental protection. Produced by the Water Research Centre, which scans over 600 primary journals, technical reports, monographs, conference papers, theses, and other printed materials.

> Coverage: 1960 to present
> Updated: Biweekly

Chemical Engineering and Biotechnology Abstracts

Half of this database covers theoretical, practical, and commercial aspects of chemical engineering, as well as chemical aspects of processing, safety, and the environment. The other half covers process and reaction engineering, measurement and process control, environmental protection and safety, plant design, and equipment used in chemical engineering and biotechnology.

More than 400 of the world's major primary chemical and process engineering journals are scanned to compile the database.

> Coverage: 1971 to present
> Updated: Monthly

Chemical Safety NewsBase

Covers a wide range of information on the health and safety effects of hazardous chemicals encountered by employees in industry and laboratories. Subjects covered include: safety precautions, chemical hazards, biological hazards, new legislation, plant and laboratory design, waste management, emergency planning, risk analysis, labeling, and storage practices. Over 200 journals, books, audiovisual materials, press releases, and conference proceedings from around the world are scanned each month for inclusion in the database.

> Coverage: 1981 to present
> Updated: Monthly

Current Awareness in Biological Sciences

Provides bibliographic information on the entire field of biological sciences. Subjects covered include biochemistry, clinical chemistry, cell biology, genetics, microbiology, ecology, plant science, pharmacology, physiology, immunology, toxicology, cancer research, neuroscience, developmental biology, endocrinology, molecular biology, and environmental sciences.

> Coverage: 1983 to present
> Updated: Monthly

Electric Power Industry Abstracts

Provides access to literature on electric power plants and related facilities. Topics include environmental effects of electric power plants and associated transmission lines; power plant siting methodologies; fuel transportation; storage, and use; licensing and permit data; energy resources; monitoring programs; safety and risk management; uranium enrichment; coastal zone management plans; waste disposal facilities; and land-use studies.

> Coverage: 1975 to 1983
> Updated: 5 times a year

ENERGYLINE

Over 70,000 journals—as well as reports, surveys, monographs, newspapers, conference proceedings, and irregular serials—are screened to provide comprehensive coverage of energy information.

> Coverage: 1971 to present
> Updated: Monthly

Enviroline

Covers air, land, and water quality; environmental health, and resource management. Source documents include journals, government reports and documents, monographs, conference papers, newspapers, the *Federal Register*, and some films.

> Coverage: 1971 to present
> Updated: Monthly

Forest

Covers worldwide literature pertinent to the wood products industry, from harvesting the standing tree through marketing the final product. Source materials include technical journals, government publications, patents, trade journals, abstract bulletins, and monographs.

> Coverage: 1947 to 1987
> Updated: Bimonthly

GeoMechanics Abstracts

Covers published literature on the mechanical performance of geological materials relevant to the extraction of raw materials from the earth and the construction of civil engineering projects and facilities for energy generation. Subjects covered include rock and soil mechanics, properties of geological materials, engineering geology, hydrology, mining, tunneling, foundation engineering, rock breakage and excavation, waste disposal, site and laboratory investigations, and analysis and design methods.

> Coverage: 1977 to present; some pre-1977 material
> Updated: Bimonthly

Georef

Covers geosciences literature, including 3000 journals plus books, conference proceedings, government documents, maps, and theses. Subjects include geology, economic geology, engineering environment geology, geochemistry, geochronology, geomorphology, igneous and metamorphic petrology, solid earth physics, and stratigraphy.

> Coverage: 1985 to present
> Updated: Monthly

ISTP Search

Covers proceedings published internationally in journals, journal supplements, serials, and monographs. This multidisciplinary file includes the following major topic areas: agriculture, applied sciences, biology, chemical sciences, clinical medicine, engineering, environmental sciences, life sciences, mathematics, physical sciences and technology.

> Coverage: 1982 to present
> Updated: Monthly

NTIS

Covers U.S. government-sponsored research by hundreds of federal agencies, and their contractors and grantees. Includes technical reports, reprints, computer software and data files, subscriptions and bibliographies. Its multidisciplinary scope includes administration and management, agriculture and food, business and economics, energy, environmental health, biomedical science, materials science, mathematical science, physics, and space technology. An increasing portion of the database consists of unpublished material originating outside the United States and obtained through international agreements.

> Coverage: 1964 to present
> Updated: Biweekly

Paper, Printing, Packaging, and Nonwovens Abstracts.

Coverage of the world's literature on all aspects of paper, pulp, nonwovens, printing, and packaging. Subjects covered include raw materials and additives, processes, machinery, equipment, pulp and paper mills, printing works, pulping, nonwovens manufacturing, types of packaging, printing special materials, packaging special products, specialty papers, testing and quality control, disposal and recycling, tamper-resistant packaging, transport of hazardous materials, and business and economic issues.

> Coverage: 1975 to present
> Updated: Bimonthly

PESTDOC

Covers worldwide literature on insecticides, herbicides, fungicides, molluscicides, and rodenticides as it relates to analysis, biochemistry, chemistry, and toxicology. Sources include journal articles, conference proceedings, and research reports.

> Coverage: 1968 to present
> Updated: Quarterly

Power

Consists of catalog records for books, monographs, proceedings, journals, and other material in the collections of the Energy Library, U.S. Department of Energy. The collection is particularly strong in general works on energy, physical and environmental sciences, technology, economics, renewable energy resources, and water resources. The library brings together significant government agency collections, including ones held by the Atomic Energy Commission, the Energy Research and Development Administration, the Federal Power Commission, and the Federal Energy Administration.

> Coverage: 1950s to present
> Updated: Quarterly

Safety Science Abstracts

Covers the broad interdisciplinary science of safety—identifying, evaluating, and eliminating or controlling hazards. Safety Scientific Abstracts focuses on six major areas: general safety, industrial and occupational safety, transportation safety, aviation and aerospace safety, environmental and ecological safety, and medical safety. Within those areas, the database covers liability information and phenomena that directly or indirectly threaten humanity, the environment, or the technology upon which people depend. The phenomena include pollution, waste, drugs, epidemics, natural disasters, fire, radiation, pesticides, genetics, toxicology, injuries, and diseases.

> Coverage: 1981 to present
> Updated: Quarterly

Standard Pesticide File

Companion file to PESTDOC. Lists approximately 3,900 known pesticides and other common compounds, including the full name and the standard registry name for each classification of standard activities (if any), chemical substructure terms, chemical ring codes, and other codes.

> Coverage: Current
> Updated: Periodically

Tropical Agriculture

Covers worldwide literature on tropical and subtropical agriculture, including crop production, crop protection, fertilizers and soils, plant nutrition, agricultural techniques, crop processing and storage, agroforestry, farming systems research and development, and environmentally sound agricultural practices.

> Coverage: 1975 to present
> Updated: Quarterly

Specialized Databases

The following specialized databases provide you with a wide choice of bibliographic services, technical information, reports, and human resources. The majority of the services are available for a fee, although a few are free. Prices range from a few dollars to hundreds of dollars per hour of use. Many of the databases are available through the commercial research databases referenced in this chapter but are mentioned here because, by dialing into them directly, you may be able to get the same information at a lower cost.

Chemical Information System

Chemical Information Systems, Inc., is a collection of chemical information databases that are searchable for a fee. Each database is a stand-alone system dealing with topics such as chemistry, toxicology, and environmental pollution.

> Cost: Each database has its own price structure, with prices ranging from $30 to $95 per hour. A $300 annual fee includes a user guide. Telecommunications charges are additional.
> Contact: Call 800/247-8737

Chemical Referral Center

The Chemical Referral Center (CRC) is a public service provided by the Chemical Manufacturers Association, a trade group representing more than 90% percent of the production of industrial chemicals in North America. CRC provides health and safety information about chemicals and chemical production to the public, transportation workers, and others who use chemicals.

To use the database, you call a toll-free number and request information about a particular chemical. CRC searches the database, and gives you the address and phone number of the manufacturer and the name of a contact there. That contact person provides the specific health and safety information about the chemical.

> Cost: Free
> Contact: Call 800/CMA-8200 between 9 a.m. and 6 p.m. Monday through Friday

Current Contents on Diskette

Current Contents on Diskette is a computer version of *Current Contents*, a weekly bibliographical indexing service from the Institute for Scientific Information. You can subscribe to the disk versions of Life Sciences; Physical, Chemical and Earth Sciences; Clinical Medicine; Agriculture, Biology and Environmental Sciences; Engineering, Technology and Applied Sciences; and Social and Behavioral Sciences. Disk versions are available for IBM PC compatibles, the NEC 9800 series, and the Apple Macintosh.

Searching is easy, and a comprehensive user manual is included. You can order a hard copy of an article you want; it will be sent within 48 hours.

> Cost: Costs run from $320 per year without abstracts to $795 per year with abstracts.
> Contact: You can get a free four-week trial subscription by calling 800/336-4474 or writing to the Institute for Scientific Information, Attention: Fulfillment Services, 3501 Market St., Philadelphia, PA 19104.

Cyclopean Gateway Service

BT Tymnet, provider of the Tymnet communications network, recently developed a menu-driven online service that gives you access to more than 850 databases from 13 worldwide information services. Cyclopean Gateway Service (CGS) is a customized version of the EasyNet Knowledge Gateway, operated by Telebase Systems, Inc. Tymnet provides the service through its Dialcom network.

Subject areas include science and technology, medicine, health, law, education, business, and the arts. News is culled from domestic and international magazines, newspapers, and other printed sources. Databases such as BRS and news services such as UPI, AP, and Reuters, are available through CGS.

Cost: Several rate structures and charges are available, with minimums of $25 per month per mailbox, a $15 per month basic fee, a $25 one-time registration fee, and hourly connect rates of $6.50 non-prime time, plus kilobyte and telecommunications charges for using the Tymnet network.
Contact: BT Tymnet at 408/922-0250.

Emergency Management Information Center

The National Emergency Training Center Learning Resource Center is a campus library for students attending the National Fire Academy, the Emergency Management Institute, and other training and education programs sponsored by the Federal Emergency Management Agency (FEMA). The library contains more than 40,000 books, reports, magazines and audiovisual materials on natural and technological disasters. Of interest to environmentalists is information on hazardous materials and nuclear incidents. Many of the resources can be borrowed through interlibrary loan. Case studies of incidents such as the accident at Three Mile Island are also available.

You call a toll-free number to request information such as an organization number or address or a publication source and price. Literature searches and complete bibliographies are prepared for you.

Cost: Free
Contact: EMIC at 800/638-1821

Firedoc

Firedoc is a computer-based bibliographic database on fire research containing over 19,000 citations covering all aspects of fire science. The database is a good source for studies on fire ecology, forestry, and the environmental effects of fire. Many fire research software programs are available on this BBS. Firedoc is available 23 hours a day Monday through Friday and all day Saturday and Sunday. It requires software that can emulate a Televideo 950 computer.

> Cost: Free
> Contact: To get a user manual and a password call Nora Jason at 301/975-6862. To learn more about Firedoc, call the Center for Fire Research Computer BBS with your modem at 301/921-6302.

GenBank

GenBank, from IntelliGenectics, is a commercial online service for molecular biologists. Several searchable databases are available, including GenBank (a nucleic acid sequence database, updated quarterly), EMBL (European Molecular Biology Laboratory's database of nucleic acid sequences, updated quarterly), GenPept (a database of peptide sequences derived by the automatic translation of annotated coding regions of entries in the current GenBank release, updated quarterly), SWISS-PROT (an annotated database of protein sequences including ones adapted from PIR, as well as many additional sequences translated from EMBL/GenBank), and GenBank Software Clearinghouse (information about molecular biology software products).

GenBank can be accessed through SprintNet and through the Internet as a telnet procedure (telnet genbank.bio.net). Recent updates of GenBank and EMBL can be received by anonymous ftp or Usenet news (bionet.molbio.genbank.updates).

> Cost: Various pricing schemes are available, but all allow free seven-minute sessions to download single

> database entries.
> Contact: IntelliGenectics at 415/962-7364.

Germplasm Resources Information Network

The Germplasm Resources Information Network (GRIN) is a centralized online database that provides scientists with information regarding plant genetics resources within the United States. It is managed by the National Plant Germplasm System Database Management Unit of the National Germplasm Resources Laboratory, part of the USDA. GRIN is a central repository of information concerning both major and minor aspects of plant genetics resources.

> Cost: Free
> Contact: To gain access to the system, write to the Database Manager, USDA GRIN, DBMU, Room 130, Building 001, BARC-West, Beltsville, MD 20705. Call 301/344-1666 for more information.

National Library of Medicine

The National Library of Medicine online databases are geared for researchers in medicine and biology. MEDLARS, a biological database set, and TOXNET, a toxicological database set, are available 24 hours a day. More than 3,500 journals are covered in the databases. The National Library of Medicine has Macintosh and IBM versions of its easy-to-use search software called Grateful Med.

TOXNET, or Toxicology Data Network, contains a number of useful databases. These include the Hazardous Substance Databank (HSDB), the Integrated Risk Information System (IRIS), the Toxic Release Inventory (TRI), and the Registry of Toxic Effects of Chemical Substances (RTECS). The Hazardous

Substance Database contains a wealth of information about individual chemicals and chemical compounds: their health and safety aspects, transport, health effects, physical chemical properties, safety hazards, firefighting tips, typical environmental concentrations, sources, and manufacturers. It also includes a list of synonyms for each chemical, so you don't have to know its exact name.

The Integrated Risk Information System (IRIS) is an official EPA database. The Agency-approved numbers used in risk assessments are here.

Though somewhat inaccurate, the Toxic Release Inventory is a useful source of data on the release of chemicals into water, air, or soil. As part of the Community Right-to-Know Act, manufacturers must report such releases.

Finally, the Registry of Toxic Effects of Chemical Substances provides a list of federal regulations applying to each chemical.

> Cost: Each database has its own hourly charge, from $23.50 per hour to $171 per hour. There is no sign-up fee and no monthly minimum. You can have your search results printed and mailed to you for 25 cents per page.
> Contact: The MEDLARS service desk at 800/638-8480 or the National Library of Medicine, Specialized Information Services at 301/496-6531. Ask to buy the Online Services Reference Manual.

Natural Products ALERT

Natural Products ALERT (NAPALERT) is a database covering the world's literature on chemical constituents and the pharmacology of plant, microbial, and animal abstracts. It is maintained by the Program for Collaborative Research in the Pharmaceutical Sciences, College of Pharmacy, University of Illinois at Chicago.

More than 75,000 articles and books, culled from about 200 journals and abstract services, cover all aspects of natural products and the use of plants and animals. Combined, the articles contain information on more than 87,000 chemical species and more than 38,000 plant and animal species and contain more than 426,000 records on biological activity associated with those species.

> Cost: Individual requests cost $10 per question, including up to three pages of computer output with bibliography. Online access subscriptions range from $100 to $10,000.
> Contact: Requests can be mailed over BITNET to NAPALERT@UICBAL. For subscription information, call 312/996-2246.

OSHA Computerized Information System

The Occupational Safety and Health Administration (OSHA) computerized database is not open by modem to the general public but you can request that the information be searched by office staff at your local OSHA office. Database files contain a wealth of information on environmental health in the workplace.

> Cost: Free
> Contact: Call your local OSHA office to make a request, for more information call 801/ 524-5366.

PaperChase

PaperChase is an online information service providing access to more than 5 million biomedical references from the National Library of Medicine's MEDLINE database. More than 4000 journals from 1966 to the present are available for searching.

PaperChase is sponsored by Beth Israel Hospital, a major teaching hospital of the Harvard Medical School.

PaperChase is menu driven and easy to use; search criteria can be by author, title, journal, or subject. You can order full-text articles online. Software engines are available for both PCs and Macintosh. An extensive user manual is included.

> Cost: There is no subscription fee, no monthly minimum, and no start-up cost. You pay $23 per hour, plus 10 cents for each reference displayed. An average search costs around $6. Searching is available 24 hours a day, seven days a week. You can access PaperChase through the CompuServe Telecommunications Network from hundreds of cities.
>
> PaperChase offers a number of pricing programs to match your searching budget for members of your organization. For example, if you have four members with a budget of $1000 for searching, PaperChase will match that at no extra cost to you, doubling your searching time.
> Contact: Call 800/722-2075, or 617/732-4800

SCIENCEnet

SCIENCEnet is designed for scientists and researchers. It contains databases, bulletin boards, special interest boards, and e-mail with gateways to the Internet, BITNET, Janet, EARN, Easylink, Dialcom, and others. SCIENCEnet is accessible through SprintNet. Many useful databases and bulletin boards are available online, with an emphasis on oceanographic studies and marine science. Hundreds of projects (mailing lists) are listed, from Acoustic Doppler Current Profiles to Zooplankton Study Group. An online membership directory makes it easy to link up with other researchers.

> Cost: A one-time $75 fee is charged for the first mailbox, and $35 is charged for each additional mailbox. You pay a monthly fee of $15, with a connect fee of $4.20 per hour from 9 p.m. to 7 a.m. Day rates are slightly higher. Some of the databases have a surcharge and require authorization to gain access.
> Contact: Omnet, Inc., at 617/265-9230.

> *Not many appreciate the ultimate power and potential usefulness of basic knowledge accumulated by obscure, unseen investigators who, in a lifetime of intensive study, may never see any practical use for their finding but who go on seeking answers to the unknown without thought of financial or practical gain.*
>
> Eugenie Clark
> *The Lady and the Sharks*, 1969

CHAPTER 14

CD-ROM: Low-Cost Information Storage for Environmental Research

The development of compact disk–read-only memory (CD-ROM) technology over the last decade has been beneficial for scientists and environmentalists. It is imperative for researchers to have at their fingertips the most current and complete information on a given subject, and CD-ROM's storage and dissemination capabilities meet that need.

The first CD-ROM drives appeared on the market in 1985, and the first commercial CD-ROM disks appeared the same year. Today, several hundred CD-ROM titles are available covering a broad range of applications, with more than 300 offered as commercial products able to run on both IBM PC and Apple systems. The Optical Publishing Association estimates that about 420,000 CD-ROM readers are installed in libraries, schools and universities. With hardware prices coming down to $500 and even lower, the association predicts an annual growth rate of 60 percent. This increase in CD-ROM reader ownership is bound to lead to an increase in the number of available titles.

Developed on the coattails of the successful audio CD market, CD-ROM is an affordable way to collect and disseminate large volumes of information at low cost. One CD-ROM disk can contain the equivalent of 250,000 pages of information, or more than 1500 floppy disks, and the same disk can include data, graphics, and sound.

CD-ROM also offers security. Since the data is in a "read-only" format, you don't have to worry about accidentally erasing or copying over information. The data is protected by a strong polycarbonate plastic coating, and the only medium that can have an effect on the disk is a laser beam. The fear of "head crashes" common on hard disks is obsolete. And since compact disks are portable, you can carry them with you anywhere.

CD-ROM is particularly useful for lengthy reference materials such as bibliographies, scientific collections, and specialized dictionaries. While some commercial CD-ROMs disks cost more than the drive, they are nonetheless a convenient way to access and store large amounts of information. A small disk replaces a

FIGURE 14-1

CD-ROM *"search and retrieval" software—such as SilverPlatter's MacSPIRS, shown here—allows you to use Boolean searching to find records.*

wall of journals. In fact, one CD-ROM disk saves the equivalent of a ton and a half of paper, which would cost over $4000 to mail (a CD-ROM disk costs 70 cents to mail). If you wanted to transmit that amount of data over a phone by way of modem, it would take about 3 weeks at 2400 baud.

Because CDs have the ability to store such large amounts of information, it is no surprise that most of the available CD-ROM titles are collections of bibliographies already found on services like Knowledge Index or BRS/After Dark. In fact, many bibliographic retrieval companies like Dialog offer several of their databases on CD-ROM. Several database companies offer their bibliographies on an annual subscription basis, sending you quarterly or monthly updates on CD-ROM disks. Some book and magazine publishers now offer CD-ROM versions of their printed encyclopedias, reference works, and past articles, and several software companies offer "information"-based programs.

How Does CD-ROM Work?

To read a disk, a CD-ROM drive spins the disk while a laser beam scans the disk's underside, reflecting off a series of pits, or lack of them. A detector measures the strength of the reflective beam (a pit scatters the laser light, making the reflection weaker) and translates the results into data that is sent to the computer.

All CD-ROM drives come with straightforward instructions and the necessary software. Macintosh CD-ROM drives attach with a cable to the SCSI port in the back of the Mac. A few files are dropped into the Macintosh system folder and allow the Mac to read the CD-ROM's format. For IBM PC-compatibles, you must insert an interface card and MS-DOS CD-ROM extensions (software drivers).

Since speed of access to data is the most critical factor in choosing a CD-ROM player, some research into available models is in order. For an excellent review of Macintosh models, see John Rizzo's article "Most Valuable Player" in *MacUser* magazine,

March 1990. Rizzo found that the fastest models on the market were those using a Toshiba-based mechanism—the CD Technology Porta-Drive and Toshiba's own XM-3201 A1. Toshiba has similar models for the IBM PC market. In addition to speed, you might look at a machine's ability to play audio CDs, how easy it is to change the SCSI address and termination, price, and warranty.

For an excellent review of PC CD-ROM drives, see "CD-ROM Drives—Finally Up to Speed" by Mitt Jones, *CPC Magazine*, October 29, 1991, pp. 283–338.

Environmental CD-ROM Titles

The following titles are a sample of current databases available on CD-ROM for environmental research. Not all the titles are affordable for the individual user—some are over $1000 each—but they are listed for your reference, since many of them can be found in libraries.

Biological Abstracts
Developed by BIOSIS, the world's largest abstracting and indexing service for the life sciences, this CD is a basic research tool for those in the biological and biomedical fields. About 250,000 records per year are indexed. Coverage starts with the 1990 calender year. One disk, updated quarterly.

> Source: SilverPlatter Information, Inc., One Newton Executive Park, Newton Lower Falls, MA 02162
> Price: $7660 retail

Birds of America
Contains the full text and images from Audubon's original 1840 first-edition octavo set of *Birds of America*. Also contains audio bird calls.

Source: CMC Research, Inc., 7150 S.W. Hampton,
Ste. C-120, Portland, OR 97223
Price: $99

Climatedata NCDC 15 Minute Precipitation

Carries precipitation records minus the National Climatic Data Central research archival files. Coverage includes 1971 to mid-1988 for the United States and approximately 2000 other stations.

Source: EarthInfo, Inc., 90 Madison St., Ste. 200,
Denver, CO 80206
Price: $1095

Deep Sea Drilling Project (DSDP)

Contains all the computerized DSDP data files from the National Geophysical Data Center. Measurements are from more than 1000 holes drilled at 624 sites in the global ocean during 16 years of operation by the Scripps Institute of Oceanography.

Source: National Geophysical Data Center,
325 Broadway, Dept. 731, Boulder, CO 80303
Price: $90

Discovery Environmental Data

Worldwide food, demographic, and agricultural data; general environmental data from the book *World Resources 1990–91*; sample energy, economic, and trade data for the industrialized countries; and more.

Source: PEMD Education Group, Ltd., 220 Hyde St.,
San Francisco, CA 94109.
Price: $149

EconLit

Includes citations and selected abstracts from the international literature on the fields of economics since 1969. Developed by the American Economic Association, it corresponds to the *Journal of Economic Literature* and *The Index of Economics Articles* and covers journal articles, dissertations, and books, as well as chapters and articles in books and conference proceedings. Topics include economic theory and history; monetary theory and financial institutions; labor economics; international, regional, and urban economics; and other related subjects. One disk, updated quarterly.

> Source: SilverPlatter Information, Inc., One Newton Executive Park, Newton Lower Falls, MA 02162
> Price: $1600 retail

The Electronic Whole Earth Catalog

Contains many items similar to those in the printed *Whole Earth Catalog* plus practical information about tools, technology, and science.

> Source: Broderbund Software, 17 Paul Dr., San Rafael, CA 94903.
> Price: $149.95 retail

Environmental Bibliography

> Source: NISC, 3100 St. Paul St., Ste. 6, Baltimore, MD 21218.

The Excerpta Medica Library Service (Excerpta Medica Abstract Journals on CD-ROM)

In-depth coverage of all aspects of human medicine, containing the equivalent of over 40 *Excerpta Medica Abstract Journals* published between 1984 and 1987. About 150,000 abstracts per year. Also included on the two-disk set are three other abstract journals: *AIDS, Forensic Science, and Environmental Health and Pollution Control.*

> Source: SilverPlatter Information, Inc., One Newton Executive Park, Newton Lower Falls, MA 02162.
> Price: $2495 retail

Facts on File News Digest
Newspaper and magazine article abstracts from 1980 to 1988. You can search by keyword, view maps, and browse. Information can be printed but not manipulated.

> Source: 460 Park Ave. S, New York, NY 10016
> Price: $770 retail

GeoRef
The American Geological Institute's GeoRef database and the Bibliography and Index of Geology have over 1.5 million citations, many with abstracts, covering literature since 1785 on the geology of North America and literature since 1933 on the geology of the rest of the world. Over 3,000 journals in 40 languages are scanned, as are books, maps, and reports. Also included are many of the U.S. Geological Survey publications and U.S. and Canadian master's theses and doctoral dissertations. Two disks, updated quarterly.

> Source: SilverPlatter Information, Inc., One Newton Executive Park, Newton Lower Falls, MA 02162
> Price: $2400 retail

MEDLINE
The entire MEDLINE database of the U.S. National Library of Medicine from 1966 to present, with a MeSH thesaurus and full

explosion capability. The database contains bibliographic citations and abstracts of biomedical literature, includes all foreign languages and all data elements, and is fully indexed. Annual subscriptions include a tutorial and are available for individual volumes, the entire set of volumes, or any number of volumes.

> Source: SilverPlatter Information, Inc., One Newton Executive Park, Newton Lower Falls, MA 02162
> Price: $750 to $1250

The National Directory

National yellow pages of over 100,000 organizations, with toll-free and fax numbers. Complete searching capability of names and addresses.

> Source: Xiphias, 8758 Venice Blvd., Los Angeles, CA 90343
> Price: $195 retail

The New Grolier Electronic Encyclopedia

Contains the text of all 21 volumes of the *Academic American Encyclopedia*. Any word can be searched.

> Source: Grolier Electronic Publishing, Inc., Old Sherman Turnpike, Danbury, CT 06816.
> Price: $399 retail

NTIS

Bibliographic citations and abstracts of unrestricted technical reports from both U.S. and non-U.S. government-sponsored research. Compiled by the National Technical Information Service from the *Government Reports Announcements and Index*.

NTIS covers topics such as engineering, biotechnology, the environment, and the physical, biological, and social sciences. Also includes extensive information on energy from the U.S.

Department of Energy. Covers 1983 to present, with over 70,000 completed reports added annually. Two disks, updated quarterly.

> Source: SilverPlatter Information, Inc., One Newton Executive Park, Newton Lower Falls, MA 02162
> Price: $2500 retail

Place Name Index

Offers quick access to more than a million place names collected from USGS quadrangle maps.

> Source: Wayzata Technology, Inc., P.O. Box 87, 16221 Main Ave., Prior Lake, MN 55372
> Price: $795

Timetable of History: Science and Innovation

HyperCard-based, covering the beginning of the universe to 1988 and all significant scientific and technological events in between. Each topic is described in 25 words or less. Boolean search capabilities are available.

> Source: Xiphias, 8758 Venice Blvd., Los Angeles, CA 90343
> Price: $185 retail

Urban Phytonarian

Tells you how to evaluate, diagnose, and treat unhealthy plants that live in the urban environment.

> Source: Quanta Press, Inc., 2550 University Ave.,W, Ste. 245N, St. Paul, MN 55114
> Price: $149

FIGURE 14-2

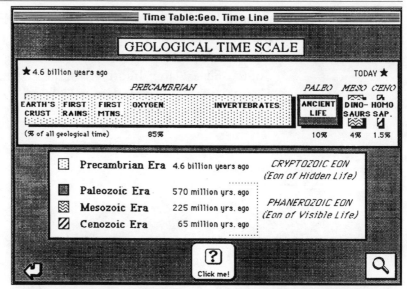

CD-ROM *technology is ideal for reference and educational use. Xiphia's Time Table of Science is an easy-to-use interactive reference work on science and technology that uses sound, animation, and graphics.*

The World Factbook
Contains geographical, political, and demographic information about more than 240 countries, based on the CIA's *World Factbook*.

> Source: Wayzata Technology, Inc., P.O. Box 807, Grand Rapids, MN 55744.

Were it left to me to decide whether we should have a government without newspapers, or newspapers without government, I should not hesitate a moment to prefer the latter.

Thomas Jefferson

CHAPTER 15

Instant Access to Environmental News

Before online services, keeping informed meant waiting for the morning paper, listening to the radio, or waiting for the local television news. Often this meant wading through articles of no interest to you or missing out on important developments in your field that were not carried by the local media. Not good options for the serious researcher! All that has dramatically changed. You can now obtain news and information tailored to your interests, 24 hours a day, and read it at your convenience.

Getting the news when it happens can make the difference between success and failure. Professionals can act quickly to gather important samples in an oil or toxic waste spill or to provide relief to earthquake victims. Medical researchers can read the latest-breaking story on AIDS or cancer research. Educators can bring up-to-date science news to their classrooms. Activists and lawyers can implement campaigns or file suits based on environmental legislation as soon as it passes.

It's easy to get overwhelmed by the volume of news available each day. Several online providers allow you to set up a personal electronic clipping service using keywords. Only those items of interest to you will be waiting when you log on, saving you a tremendous amount of search time (and mental overload). Access to the news as it happens allows you to act on, not react to, environmental issues that interest you.

This chapter provides an overview of the news services available on CompuServe, GEnie, and America Online, which you can use to create and maintain your own personal daily environmental newspaper. (Those online services are covered in detail in Part IV of this book.) This chapter also examines three commercial news and information providers: NewsNet, Inc., UPI Environmental NewsNow, and the Environment News Service.

CompuServe

Sources for CompuServe's news services include the Associated Press (AP), United Press International (UPI), Reuters, the *Washington Post,* and McGraw-Hill. You can access the news services from the CompuServe main menu.

Figure 15-1 shows the News menu of CompuServe. This

FIGURE 15-1

```
20H 1 Executive News Service (E$)
 2 NewsGrid
 3 AP Online
 4 Weather
 5 Sports
 6 The Business Wire
 7 Newspaper Library
 8 Entertainment News/Info
 9 Online Today Daily Edition
50HJEnter choice !3
```

The News menu of CompuServe. To make a selection, type the number of your choice at the prompt.

section describes the Executive News Service, AP Online, and the Newspaper Library.

Executive News Service

The Executive News Service (ENS) is an electronic clipping service that monitors AP and UPI news wires, the *Washington Post,* and *Reuters Financial Report*. As a subscriber to ENS, you specify search words or phrases for topics that interest you, and ENS "clips" news stories that contain those words or phrases and holds them in "folders" for you to retrieve and read.

Since ENS is one of CompuServe's Executive Option services, you pay a surcharge to use it. You can use the command `GO ENS` to get to the service.

AP Online

Associated Press Online allows you to review hourly the latest AP stories in a number of news categories. (See Figure 15-2.) You can access AP Online by typing `GO APO`.

You feel a unique sense of confidence and excitement knowing that you are in touch with world events at the touch of your keyboard. One particular night that I accessed AP Online, I learned about the 1990 earthquake in Nicosia moments after it

FIGURE 15-2

1 Latest News Update Hourly	7 Entertainment
2 Weather	8 Business News
3 National	9 Wall Street
4 Washington	10 Dow Jones Avg
5 World	11 Feature News
6 Political	12 History

CompuServe's AP Online allows you to review the latest news by the hour.

Newspaper Library

The Newspaper Library contains selected full-text articles from 40 newspapers across the United States. Articles are available through the library two days after they first appear in print.

In addition to the usual CompuServe connect charges, you pay $3 for each search of ten titles, and $3 to read each full-text article. You can search the database using subject keywords and Boolean search connectors like *AND*, and *OR*. The Newspaper Library can be accesed with the comand `GO NPL`

Table 15-1 shows the names of newspapers available for full-text search in the Newspaper Library.

TABLE 15-1 Newspapers for which full-text articles are available for full-text search in CompuServe's Newspaper Library

City	Newspaper	City	Newspaper
Akron (Ohio)	*Beacon Journal*	Miami (Fla.)	*Herald*
Albany (N.Y.)	*Times-Union*	New Jersey	*Record*
Allentown (Pa.)	*Morning Call*	Newark (N.J.)	*Star-Ledger*
Anchorage (Ala.)	*Daily News*	New York (N.Y.)	*Newsday*
Annapolis (Md.)	*Capital*	Orlando (Fla.)	*Sentinel*
Arizona	*Republic/ Phoenix Gazette*	Palm Beach (Fla.)	*Post*
		Philadelphia (Pa.)	*Daily News*
Atlanta (Ga.)	*Constitution/ Atlanta Journal*	Philadelphia (Pa.)	*Inquirer*
		Richmond (Va.)	*News Leader/ Richmond Times-Dispatch*
Boston (Ma.)	*Globe*		
Buffalo (N.Y.)	*News*		
Charlotte (N.C.)	*Observer*	(Colo.)	*Rocky Mountain News*
Chicago (Ill.)	*Tribune*		
Columbia (S.C.)	*State*	Sacramento (Calif.)	*Bee*
Columbus (Ohio)	*Dispatch*	San Francisco (Calif.)	*Chronicle*
Detroit (Mich.)	*Free Press*	San Jose (Calif.)	*Mercury News*
Fort Lauderdale (Fla.)	*News and Sun-Sentinel*	Seattle (Wash.)	*Post-Intelligencer*
		St. Louis (Mo.)	*Post-Dispatch*
Fresno (Calif.)	*Bee*	St. Paul (Minn.)	*Pioneer Press Dispatch*
Gary (Ind.)	*Post-Tribune*		
Houston (Tex.)	*Post*	St. Petersburg (Fla.)	*Times*
Lexington (Ky.)	*Herald-Leader*	Washington (D.C.)	*Post*
Los Angeles (Calif.)	*Daily News*	Wichita (Kans.)	*Eagle-Beacon*
Los Angeles (Calif.)	*Times*		

Keeping the Bears Alive

In 1989, Bryan Bashin, a science writer, learned of the legally-sanctioned killing of thousands of black bears each year on private timberlands in Washington, Oregon, and California. Louisiana Pacific and other timber companies justified the hunts because they prevented the bears from killing trees. The problem was, the bears regularly clawed and ate the bark of the trees each year as they emerged from hibernation. This cut into the timber companies' profits.

At first, the companies persuaded trappers hired by the federal Animal Damage Control Agency to shoot the bears at no cost to them. In some states, timber companies hired their own hunters. Finally, the companies got their State Fish and Game departments to authorize hunts on their land, so hunting permits could be sold to private citizens.

Bashin learned of this from a brief item in *Newsweek* magazine about an award given to a former trapper who had learned to prevent bears from damaging trees. Bashin searched newspaper archives on DataTimes for the man's name, and found an article on him in a Seattle newspaper.

The man had shot 1600 bears during his stint as a trapper, and had decided to look for a better solution. So, he set out pans of food for the bears for a few critical weeks in early Spring only in areas where the bears had nothing else to eat. The bears preferred the prepared food over bark, and the problem went away.

Bashin wrote articles on this for *Sierra* magazine and the Sunday supplements of the *Sacramento Bee* and the *San Francisco Examiner*. Soon afterwards, a Sacramento environmental law firm filed suit against the State of California Fish and Game Department.

> The Superior Court judge in Sacramento froze the entire state's bear hunt for one year, ruling that the State had no evidence to support the necessity of shooting bears. This saved the lives of one or two thousand bears in California.
>
> In the year or so since then, the Fish and Game Department in California has reworked its entire rationale for authorizing hunts. The Department now requires detailed Environmental Impact Statements discussing the health of the bears, and alternatives to killing them, before it will authorize hunts on timber companies' property.
>
> "Bears are still being shot for sport on public land in California, but at least now they're not being shot because they kill trees," Bashin says.
>
> Bashin can be reached by MCI mail.
>
> —Wendy Monroe

GEnie

Figure 15-3 shows the GEnie news services that can be accessed either by typing `news` at any prompt, or by typing the page number of the desired menu with the word *move*—for example, `Move 300`.

NewsGrid

NewsGrid is a real-time news service compiled from the dispatches of eight of the world's largest wire services:

- Agence France Presse   (France)
- BusinessWire   (United States)
- Deutsch Press-Agentur   (Germany)
- Agencia EFE   (Spain)
- Kyodo News Service   (Japan)

- ◆ PR Newswire (United States)
- ◆ United Press International (United States)
- ◆ Xinhua News Agency (China)

NewsGrid is broad in scope but has a strong business focus. Coverage includes world affairs, politics, and environmental issues. You can search for environmental stories using keywords such as *environment*, or *pesticides*.

NewsGrid offers two great features: It gives you the ability to correspond with journalists and other readers from around the world to give immediate feedback or to request more information. Also, NewsGrid allows you to create a personal news clipping service so you don't miss important articles in the areas you select.

If you are interested in discussing current news events with other GEnie members, NewsGrid's LiveWire NewsRoom is the place to go. In the LiveWire NewsRoom, news is displayed 24 hours a day as it comes over the wire; it's like having your own teletype machine. A unique aspect of the LiveWire NewsRoom is the dimension of "interactivity" it gives to the news. LiveWire allows you to discuss news with others online as the news rolls off the wires. Enter `news` or `Move 340` to get to NewsGrid.

FIGURE 15-3

```
GEnie NEWS    Page 300           ←——————— GEnie's News menu
  US & World News
  1. NewsGrid Headline News ←———  Provides general (including
                                  environmental) news
  2. Press Releases
  3. Personal Computer News
  4. Home Office/Small Business RT
  5. GEnie QuikNews
  6. Newsbytes News Network ←———  Provides news about the
                                  computer industry
  7. This Week In History
  8. Dow Jones News/Retrieval
Enter #, <P>revious, or <H>elp?1
```

GEnie's news service menu.

FIGURE 15-4

```
GEnie    NEWSBYTES    Page 316
   Newsbytes News Network


   1. About Newsbytes
   2. Newsbytes Instructions
   3. This Issue's Index
   4. Retrieve News by Bureau
   5. Retrieve News by Subject
   6. Retrieve News by Story Type
   7. Download an Issue of Newsbytes
   8. The Mailbag (Reader Mail)
   9. Send FEEDBACK to Newsbytes
Enter #, <P>revious, or <H>elp? 5
```
⟵ *I choose to search by subject area*

```
      Newsbytes News Network
      Subject Selections


   1. Apple
   2. IBM
   3. UNIX
   4. Trends
   5. Business
   6. Telecom
   7. General
   8. Government
Enter #, or <P>revious? 6
```
⟵ *I choose to look for articles related to telecommunications*

```
         Newsbytes News Network
      Issue Selection - Telecom


   1. This issue's stories
   2. Last issue's stories
   3. All available stories
Enter #, <P>revious or <Q>uit? 1
```
⟵ *I decide to check the most recent stories first*

```
            Newsbytes News Network
              Telecom Headlines


    1. NEW TRAVEL RESERVATION NETWORK DEBUTS
    2. MODEM KEEPS GREENPEACE IN TOUCH WITH THE WORLD
    3. HONGKONG: CELLULAR PHONE SABOTAGES TRADER
    4. FEARS OF TELEPHONE CHAOS AS CHINA GOES ITS OWN WAY ON
  STANDARDS
    5. HONG KONG ADDS MORE CHINA DESTINATIONS TO IDD CALL LIST
    6. TANDEM CREATES DOW JONES/KYODO NEWS LINK IN JAPAN
    7. NEC LAUNCHING ONLINE SERVICE FOR PC ENGINE THIS FALL
    8. NORTEL, BELL-NORTHERN CLAIM SONET FIRST

          [More articles were found in this search.]

  Enter #'s, <M>ain News Menu,
  <P>revious or <Q>uit?  2          ⟵  I choose to read the story
                                        about Greenpeace.

  MODEM KEEPS GREENPEACE IN TOUCH WITH THE WORLD
  SYDNEY, AUSTRALIA, 1990 JUN 7 (NB)—NetComm has provided the
  Australian arm of Greenpeace with an AutoModem 1234 for use
  in the organization's international communications
  activities. The modem allows Greenpeace Australia to send
  and receive updates on international happenings in the
  environmental and disarmament sphere.

                                    ⟵ The story continues, but
                                       we'll stop here.
```

A search of GEnie's Newsbytes service for stories on the environment. Comments and user input are boldfaced.

Newsbytes News Network

Newsbytes is billed as the largest independent computer industry news service in the world, with four U.S. bureaus and seven international bureaus, in London, Brussels, Toronto, Tokyo,

Hong Kong, Moscow, and Sydney. Newsbytes typically contains articles on the environment, science, and technology.

Newsbytes is carried by America Online and other providers, but GEnie is the official network for communication among Newsbytes' bureaus. To get to Newsbytes, you can type `Move 316`.

You can search Newsbytes articles by bureau, subject, or story type. Figure 15-4 shows a sample search for environmental stories.

Newsbytes also allows you to send letters and feedback to the editor, Wendy Woods (who also works on "Computer Chronicles," a television show aired on more than 180 PBS stations nationwide). Wendy responds to her mail online. This is a great way to get rare, behind-the-scenes insight into how news is made and reported. Newsbytes is also available on CD-ROM disc (1983–1991) for $129 from Wayzata Technology, 412 Pokegama Avenue, Grand Rapids, MN 55744.

America Online

America Online's news service can be reached by clicking on the News icon on the main menu or by using the keyword `news`. News stories can be read online or saved to your computer to read later.

From America Online's Environmental Forum, located in the Lifestyles and Interests department, you can download a HyperCard stack called the Environmental News Weekly Reader. Each week, the forum host fills the stack with the previous week's environmental news and makes it available for downloading in the Environmental Forum's file library.

NewsWatch

America Online's NewsWatch service allows you to use keywords to search for the day's environmental news stories (see Figure 15-5).

FIGURE 15-5

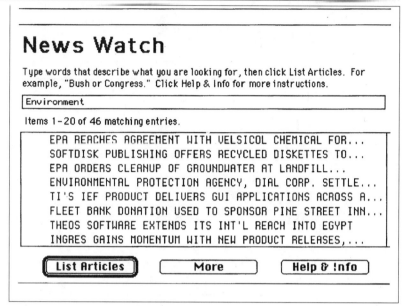

You can search America Online's NewsWatch by keyword to find the latest environmental news stories.

Newsbytes

The Newsbytes news network, already discussed in the GEnie section of this chapter, is also available on America Online. The graphic interface of America Online makes it easy to use Newsbytes and download files (see Figure 15-6).

NewsNet

NewsNet, Inc., a subscription online news search service, provides the full text of articles from more than 450 specialty newsletters and other news sources (graphs, charts and ads are not included). NewsNet was created for those who want to stay on top of information about the business world, but the service also

FIGURE 15-6

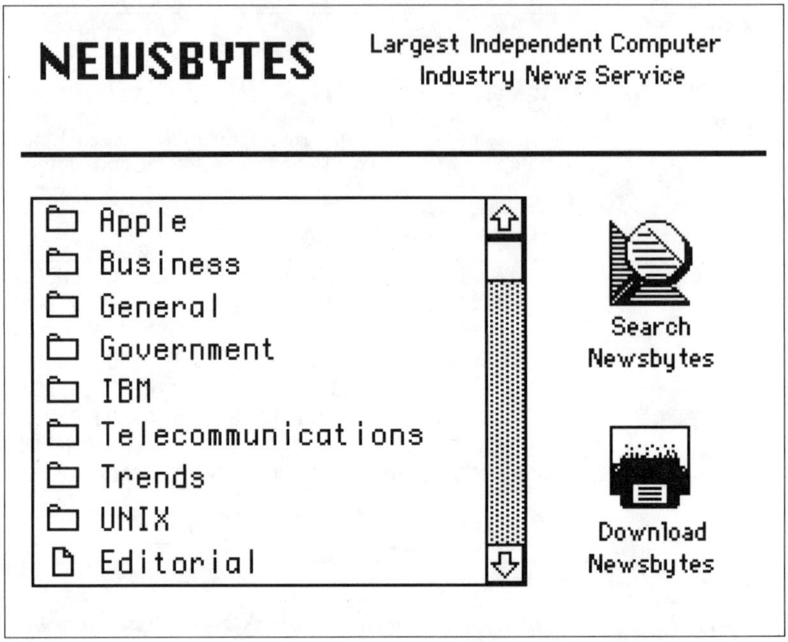

```
NEWSBYTES          Largest Independent Computer
                      Industry News Service

   📁 Apple
   📁 Business                              [Search
   📁 General                               Newsbytes]
   📁 Government
   📁 IBM
   📁 Telecommunications
   📁 Trends                                [Download
   📁 UNIX                                  Newsbytes]
   📄 Editorial
```

Newsbytes on America Online can be searched by keyword; you can also download all the articles in categories that interest you.

provides informative articles on the environment, medicine, and science fields.

A NewsNet feature called NewsFlash lets you create your own clipping service. It automatically reports to you any new reference material on a chosen topic when you sign on.

NewsNet is an ideal source for the latest articles on pressing environmental issues such as nuclear, hazardous, and medical waste; groundwater and air pollution; energy; health care; federal environmental legislation and testimony; biotechnology; and research and development.

How to Get Online

To get information about how to subscribe and about the services offered, call 800/345-1301.

Cost

NewsNet offers various payment options. Annual, half-year, and monthly subscription fees are $120, $75, and $15 respectively and include a regularly updated, comprehensive user manual.

If you have an annual subscription, online charges are $60 per hour for 300 to 1200 baud and $90 per hour for 2400 baud. On top of those charges, you will pay "read charges" to the publisher of titles you access. The NewsFlash clipping service costs 50 cents for every "hit" on your area of interest. Carefully preparing your searches before you sign on to NewsNet can help keep your online costs to a minimum.

Organization

NewsNet offers four categories of resources: newsletters, directories, wire services, and gateways. Newsletters are available in more than 30 fields, including aerospace, biotechnology, chemistry energy, the environment, government and regulation, health and hospitals, medicine, and politics. The environment category contains eleven publications, from *Air/Water Pollution Report* to *WasteTreatment Technology News*.

Wire services include Associated Press's Datastream business news wire, the Business Wire, Catholic News Service, UPI, and the Chinese Xinhua English Language News Service, as well as other foreign news services. Directories such as the NASA Software Directory, the Local Area Networking Sourcebook, and the Telephone Industry Directory are updated periodically. Finally, gateway services connect you to services such as TRW, a national credit profile company; the Official Airline Guide, Electronic Edition; and VU/QUOTE Stock and Commodity Quotes.

Getting Around

NewsNet is command driven. Each publication and service is represented by a code ID that you use for searching. Codes are grouped according to subject area. For example, all the environmental publications start with the code EV; EV30 is the code for Medical Waste News, EV25 is for Environment Week, and so on. Searching is relatively easy, but the user manual is invaluable.

Kodak Takes Responsibility

John Tokofsky (not his real name), a Philadelphia environmental engineer, was browsing a pollution prevention conference when he spotted a promotional Kodak booth. Raising an eyebrow, he could not resist asking the nearby representative whether Kodak's exhibit of its Fun Saver disposable camera at an Earth Day conference was appropriate. Tokofsky may have used a Fun Saver camera the weekend before, but he could not imagine the rationale for marketing it as an "environmentally friendly" product.

The rep stiffened at his question. "Those cameras are recycleable!" she said. But, she added, show participants had been glaring at her booth. Some of them had muttered snidely about "disposable cameras for sale on Earth Day."

Several days later at work, Tokofsky scanned the new press releases on the Business Wire, carried by NewsNet. Here was a brand-new one from Kodak, inaugurating a recycling program for Fun Saver disposable cameras.

This program hearkened back to the early days of photography, Tokofsky read. In those days, Kodak accepted customers' cameras containing exposed film, processed the film, and returned the cameras with fresh film.

"What timing," Tokofsky thought. "Did the Kodak rep at the show speak with one of the heads in Marketing? Perhaps he or she wants to see an end to the landfills piled with throwaway plastic cameras?"

He captured and printed out the press release, and headed to his local photo store with his Fun Saver camera full of exposed film. He saw Kodak's cardboard display on the counter featuring photos of happy people, but no mention of a recycling plan. Setting his cardboard-wrapped plastic box on the counter, he asked the clerk if his store could recycle it.

The clerk showed Tokofsky a heap of empty camera shells in a back room. "I'm collecting them," he said. "Kodak will accept them, but it's not practical to send them back until I've got thousands."

Tokofsky handed Kodak's press release to the clerk and urged him to call the contact number on the bottom.

When Tokofsky returned a few days later, he asked the clerk whether he had called the contact number. The clerk said the Kodak marketing person he spoke with reassured him the cameras actually were recycleable and promised to send a truck out to pick them up.

Kodak currently promotes the recycling program for retailers, offers recycling seminars to dealers of the plastic cameras, and has cut the minimum number of camera shells per return shipment to 50. The company also offers a 5 cent rebate for each one. A marketing spokesman says cagily that perhaps half the cameras are being recycled, and the proportion is growing.

Tokofsky maintains Fun Saver camera users must press Kodak to live up to its good intentions. Perhaps a few phone calls to the 800-number and inquiries at stores selling the cameras are enough, he says.

He also feels a non-antagonistic attitude is key to winning over corporate decision-makers. "I like the product, and I hope to soon see recycled plastic content listed on the packaging. Maybe Kodak's leadership will be followed by music CD makers looking for a way to handle the 'long box' issue."

—Wendy Monroe

Selected NewsNet Environmental Publications

The following is a list of publications of environmental interest carried by NewsNet. New publications are added all the time, so call NewsNet for the latest offerings.

Biotechnology

Applied Genetics News
You can use this publication to monitor the business and science of biotechnology.

> Code: BT03
> Publisher: Business Communications Co., Inc.
> Frequency: Monthly
> Earliest NewsNet Issue: October 1, 1989

Bioprocessing Technology
Emphasizes the industrial production of chemicals and energy, including the conversion of biomaterials via fermentation and biodegradation.

> Code: BT01
> Publisher: Technical Insights, Inc.
> Frequency: Monthly
> Earliest NewsNet Issue: January 1, 1989

Biotech Business
You can find news and information on biotechnology products, developments, and companies here. Biotech highlights marketing and investment opportunities and reports on finances, management, company mergers, clinical drug trials, and scientific breakthroughs.

> Code: BT06

Publisher: Worldwide Videotex
Frequency: Monthly
Earliest NewsNet Issue: July 1, 1988

Genetic Technology News

Read this magazine if you are interested in genetic engineering and its uses in the chemical, pharmaceutical, food processing, and energy industries.

Code: BT02
Publisher: Technical Insights Inc.
Frequency: Monthly
Earliest NewsNet Issue: January 1, 1989

Chemical

Hazardous Waste News

Covers federal and state hazardous waste management activities, including legislation, regulations, grants and contracts, and meetings.

Code: CH10
Publisher: Business Publishers, Inc.
Frequency: Weekly
Earliest NewsNet Issue: January 4, 1982

State Regulation Report: Toxics

A good place to keep current on toxic substances control and hazardous waste management at the state level, and on national developments from a state perspective.

Code: CH11
Publisher: Business Publishers Inc.
Frequency: Biweekly

Earliest NewsNet Issue: March 15, 1985

Toxic Materials News
Covers in detail legislation, litigation, and regulations relative to the Toxic Substances Control Act.

> Code: CH12
> Publisher: Business Publishers, Inc.
> Frequency: Weekly
> Earliest NewsNet Issue: January 6, 1982

Energy

Energy Conservation News
Focuses on the technology and economics of energy conservation at industrial, institutional, and commercial facilities.

> Code: EY59
> Publisher: Business Communications Co., Inc.
> Frequency: Monthly
> Earliest NewsNet Issue: September 1, 1989

The Energy Daily
Publishes information on the energy industry and its regulators.

> Code: EY57
> Publisher: King Publishing Group, Inc.
> Frequency: Daily
> Earliest NewsNet Issue: April 10, 1989

The Energy Report
News on government action and environmental issues related to energy sources such as oil, natural gas, coal, nuclear, and alternative fuels.

> Code: EY50
> Publisher: Pasha Publications, Inc.
> Frequency: Weekly
> Earliest NewsNet Issue: August 28, 1989

HydroWire
Reports on the North American hydroelectric industry.

> Code: EY53
> Publisher: HCI Publications
> Frequency: Biweekly
> Earliest NewsNet Issue: November 30, 1987

Offshore Gas Report
Focuses on review and analysis of industry trends and government actions affecting the offshore gas market.

> Code: EV56
> Publisher: Atlantic Information Services, Inc.
> Frequency: Biweekly
> Earliest NewsNet Issue: November 15, 1988

Environment

Air/Water Pollution Report
Covers state, local, and federal environmental legislation, regulation, and litigation.

> Code: EV10
> Publisher: Business Publishers, Inc.
> Frequency: Weekly
> Earliest NewsNet Issue: January 4, 1982

Asbestos Control Report

The only independent publication to focus exclusively on the asbestos control industry. Contains information on control techniques, work site health and safety, federal standards, waste disposal, and sampling and analysis technology.

> Code: EV27
> Publisher: Business Publishers, Inc.
> Frequency: Biweekly
> Earliest NewsNet Issue: Decembet 7, 1989

Environment Week

Reports on issues ranging from acid rain and the greenhouse effect to nuclear and hazardous waste disposal.

> Code: EV25
> Publisher: King Communications Group, Inc.
> Frequency: Weekly
> Earliest NewsNet Issue: February 23, 1989

Greenhouse Effect Report

Reports on congressional, regulatory, business, technological, and international actions on global warming and the greenhouse effect.

> Code: EV29
> Publisher: Business Publishers, Inc.
> Frequency: Monthly
> Earliest NewsNet Issue: December 1, 1989

Indoor Air Quality Update

Offers practical solutions to indoor air problems. Includes research findings and alternative building and design options.

> Code: EV33
> Publisher: High Tech Publishing Co.

Frequency: Monthly
Earliest NewsNet Issue: October 1, 1985

Multinational Environmental Outlook

An important publication that covers environmental news from Europe, Asia, Africa, and Latin America.

Code: EV01
Publisher: Business Publishers, Inc.
Frequency: Biweekly
Earliest NewsNet Issue: January 5, 1981

Oil Spill Intelligence Report

The only weekly source of global information on oil spill cleanup, control, and prevention.

Code: EV32
Publisher: Cutter Information Group
Frequency: Weekly
Earliest NewsNet Issue: May 3, 1990

Superfund

A regular update on Superfund cleanup and litigation, federal and state hazardous waste cleanup programs, remedial technologies, and EPA actions.

Code: EV22
Publisher: Pasha Publications
Frequency: Biweekly
Earliest NewsNet Issue: August 3, 1990

Law

Industrial Health and Hazards Update
An excellent publication covering industrial health and hazards, toxicity of produced materiels and products, diseases dealing with occupational exposures, exposure limits, and regulations.

> Code: LA04
> Publisher: Merton Allen Associates
> Frequency: Monthly
> Earliest NewsNet Issue: April 1, 1984

UPI Environmental NewsNow

Environmentalists can have their own news feed directly to their home or office via UPI, the world's largest independent news and information service. At $495 a month, UPI Environmental NewsNow is not for everyone; however, a subscription to the service is within the means of larger environmental organizations and libraries.

UPI Environmental NewsNow brings you up-to-the-minute environmental news from UPI's 50 state wires and its national/international wire. Other sources include Agenica EFE, Business-Wire, Comtex, Invest/Net, the Xinhua News Agency, and the Kyodo News Service.

Additionally, more than two dozen environmental newsletters published by Business Publishers, Inc., are available, including: Air Toxics Report, Air/Water Pollution Report, Asbestos Control Report, Ecology USA, Clean Water Report, Greenhouse Effect Report, Hazardous Waste News, Medical Waste News, Multinational Environmental Outlook, Noise Regulation Report, Nuclear Waste News, Report on Defense Plant Waste, Environmental Health Letter, Solid Waste Report, Toxic Material News, Occupational Health and Safety Letter, and State Regulation Report.

How to Get Online

Call United Press International at 800/UPI-8870 for subscription information.

UPI Environmental NewsNow is available only for IBM PCs and PC compatibles. No satellite dish is required in cities that have FM sideband transmission—Washington, D.C., Philadelphia, Atlanta, New York City, Miami, Boston, Chicago, Houston, Los Angeles, San Francisco, and Salt Lake City. Subscribers in other areas receive a .9 meter dish installed for reception.

Cost

$495 a month for a single user includes installation, software, and fine-tuning. A contract and a deposit are required for the first 90 days.

Environment News Service

The Environment News Service (ENS) is the world's first international news agency dedicated to the independent gathering and reporting of environmental news. ENS delivers environmental news daily—by means of electronic networks, computer links, and fax—to organizations, newspapers, magazines, and online services.

ENS news is gathered by a global network of correspondents in world capitals and remote locations. Some ENS stories may appear in commercial newspapers, magazines, and news services, but since many of the stories do not make the mainstream press, it is advantageous to subscribe directly to ENS. Its journalists, lawyers, scientists, engineers, consultants, and even explorers guarantee subscribers exclusive, frontline environmental news. By actively communicating with more than 4000 environmental experts and organizations, ENS provides a worldwide perspective on environmental issues.

ENS has set up an area in the America Online Environmental Forum through which you can correspond with the editor and download daily news. ENS is also available on EcoNet, as well as in print, video, and audio formats that include all rights to reprint and broadcast.

For an annual subscription fee, you can have ENS's daily E-SHEET faxed directly to your home or office five days a week. ENS also scans all major international news wires and computer networks to compile a summary called WorldScan, which is published along with ENS's original news items.

How to Subscribe

Call 604/732-4000 or fax 604/732-4400 for information, or write the Environment News Service, at 3505 W. 15th Ave., Vancouver, B.C. V6R 2Z3 Canada.

Cost

Subscription to daily fax service is $320 per year.

APPENDIX A

Selected Communications Software

The following is a representative sample of communications software programs that you can use to sign on to many of the bulletin boards described in this book. Many of those mentioned here are available free or at low cost. If you join a computer user group, fellow members are usually happy to recommend their favorite free or shareware communications programs. When selecting a program, be sure it includes two or more error-checking protocols, such as Xmodem and Ymodem, and that the baud rate, parity, and stop bits can be changed easily.

Amiga

JRComm Jack Radigan, P.O. Box 968, Mays Landing, NJ 08330. Shareware.

Apple II

Apple Access II Apple Computer, Inc., 20525 Mariani Ave., Cupertino, CA 95014.

ASCII Pro and ASCII Express United Software Industries, 8399 Topanga Canyon Blvd., Ste. 200, Canoga Park, CA 91304.

Hayes Terminal Program Micromodem II Hayes Microcomputer Products, 705 Westech Dr., Norcross, GA 30092. Compatible with CPM, DOS 3.3, and PASCAL systems.

ModemWorks Morgan Davis Group, 10079 Nuerto Ln., Rancho San Diego, CA 92078.

Mousewrite Roger Wagner Publishing, Inc., 1050 Pioneer Way, Ste. P, El Cajon, CA 92020.

Softerm 2 Softronics, Inc., 7899 Lexington Dr., Ste. 210, Colorado Springs, CO 80920.

Term-Talk Computer Aids Corp., 124 W. Washington, Ste. 220, Fort Wayne, IN 46802.

Atari

Flash Antic Software, 544 Second St., San Francisco, CA 94107.

Commodore

Mcterm Ariel Software, Ltd., 210 N. Bassett, Madison, WI 53703.

CPM Machines

ASCII Pro and ASCII Express United Software Industries, 8399 Topanga Canyon Blvd., Ste. 200, Canoga Park, CA 91304.

Modcom Holliday Software, 4807 Arlene St., San Diego, CA 92117.

Heath/Zenith

ZLYNK/II Software Wizardry, Inc., 8 Cherokee Dr., St. Peters, MO 63376.

IBM and IBM Compatibles/DOS

ASCII Pro and ASCII Express United Software Industries, 8399 Topanga Canyon Blvd., Ste. #200, Canoga Park, CA 91304.

BackComm RML Associates/Ackcomm Software, 991-C Lomas Sante Fe Dr., Ste. 223, Solana Beach, CA 92075.

Carbon Copy Plus Meridian Technology, Inc., 7 Corporate Park, Ste. 100, Irvine, CA 92714.

Crosstalk Digital Communications Associates, 1000 Holcomb Woods Pkwy., Roswell, GA 30076.

HyperACCESS Hilgraeve, Inc., P.O. Box 941, Monroe, MI 48161.

Intelliterm Microcorp, 913 Walnut St., Philadelphia, PA 19107.

Lync Communications Software 5.0 Norton-Lambert Corp., P.O. Box 4085, Santa Barbara, CA 93140.

PC-Talk III The Headlands Press, P.O. Box 862, Tiburon, CA 94920.

Procomm Datastorm Technologies, Inc., P.O. Box 1471, Columbia, MO 65205.

Q-Modem The Forbin Project, P.O. Box 702, Cedar Falls, IA 50613.

Telix PTEL, P.O. Box 130, West Hill, Ontario, M1E 4R4 Canada.

Macintosh

Freeterm Available free from user groups and online services.

Kermit Available for downloading from online services and bulletin boards. Columbia University freeware.

MacKnowledge Prometheus Products, 7225 S.W. Bonita, Tigard, OR 97223.

MacTerminal Apple Computer, Inc., 20525 Mariani Ave., Cupertino, CA 95014.

MicroPhone Software Ventures, 2907 Claremont Ave., Ste. 220, Berkeley, CA 94705.

Smartcom Hayes Microcomputer Products, 705 Westech Dr., Norcross, GA 30092.

Termworks Available for downloading from online services and bulletin boards. Shareware.

VersaTerm Synergy Software, 2457 Perkiomen Ave., Reading, PA 19606.

White Knight The Freesoft Co., 150 Hickory Dr., Beaver Falls, PA 15010.

Zterm David P. Alverson, 5635 Cross Creek Ct., Mason, OH 45040. Available for downloading from online services and bulletin boards.

Tandy

Ultra Term United Software Associates, 734 Flamingo Wy., North Palm Beach, FL 33408.

APPENDIX B

Internet Mailing Lists

This appendix lists environmental discussion lists available through Internet electronic mail. A complete list of all Internet mailing lists is available to download from Internet by means of anonymous ftp (use the user ID `anonymous`, and the password `guest`) from host ftp.nisc.sri.com (192.33.33.53) in directory netinfo. The path name of the file is netinfo/interest-groups.

To receive the file via e-mail, send a request to `interest-groups-request@nisc.sri.com`

Send a message to the same address to submit new descriptions of mailing lists, update existing information, or delete old mailing list information from the List-of-Lists.

Agriculture

AG-EXP-L%NDSUVM1.BITNET@CUNYVM.CUNY.EDU
Participants in this mailing list discuss the use of expert systems in agricultural production and management. The list is aimed at practitioners, extension personnel and experiment station researchers in the land grant system. If you are a subscriber to BITNET, EARN, or NetNorth, you can join the list by sending the

Listserv SUB command with your name—for example, `SEND LISTSERV@NDSUVM1 SUB AG-EXP-L John Doe` or `TELL LISTSERV AT NDSUVM1 SUB AG-EXP-L Jane Doe`. To be removed from the list, enter `SEND LISTSERV@NDSUVM1 SIGNOFF AG-EXP-L` or `TELL LISTSERV AT NDSUVM1 SIGNOFF AG-EXP-L`.

Those without interactive access may send the Listserv command portion of the above lines as the first text line of a message. For example, `SUB AG-EXP-L John Doe` would be the only line in the body (text) of mail sent to `LISTSERV@NDSUVM1`. Monthly public logs of mail to AG-EXP-L are kept on LISTSERV for a few months. For a list of files send the command `Index AG-EXP-L` to `LISTSERV%NDSUVM1.BITNET@CUNYVM.CUNY.EDU`.

> Coordinator: Sandy Sprafka
> <NU020746%NDSUVM1.BITNET@CUNYVM.CUNY.EDU>

Animal Rights

ANIMAL-RIGHTS@CS.ODU.EDU
This unmoderated list is for the discussion of animal rights. The list provides students, researchers, and activists with a forum for discussing issues such as consumer product testing; cruelty-free products; medical testing; hunting, trapping and fishing; animals in entertainment; factory farming; fur; vegetarianism; and Christian and other perspectives.

All requests to be added to or deleted from the list should be sent via the Internet to `Animal-Rights-Request@[XANTH.]CS.ODU.EDU` or via UUCP to `Animal-Rights-Request@xanth.uucp`.

> Coordinator: Chip Roberson
> <csrobe@CS.WM.EDU>

Biological Sciences/Natural History

BEE-L%ALBNYVM1.BITNET@CUNYVM.CUNY.EDU
BEE-L is a mailing list for the discussion of research and the exchange of information concerning the biology of bees. Topics include sociobiology, behavior, ecology, adaptation and evolution, genetics, taxonomy, physiology, pollination, and flower nectar and pollen production.

To subscribe, send the command `SUB BEE-L` plus your name—for example, `SUB BEE-L Jane Doe`, to `LISTSERV@ALBNYVM1` via mail or interactive message. If you do not have access to BITNET you can subscribe to BEE-L by sending the text `SUB BEE-L` plus your full name in the body of a message to `LISTSERV%ALBNYVM1.BITNET@CUNYVM.CUNY.EDU`.

> Coordinator: Mary Jo Orzech
> <MJO%BROCK1P.BITNET@CUNYVM.CUNY.EDU>

BIO-NAUT%IRLEARN.BITNET@VM1.NODAK.EDU
BIO-NAUT on LISTSERV@IRLEARN.BITNET
This mailing list was established at IRLEARN to enable life scientists to communicate with each other. The list has three main functions:

- It provides the network addresses of researchers in the biological sciences.

- It allows subscribers to upload small databases of e-mail addresses related to a particular field onto BIO-NAUT for the benefit of others. Special interest groups such as numerical taxonomists or culture collections are welcome.

- It compiles information about each subscriber and his or her main interests.

To subscribe to the BIOSCI BIO-NAUTS distribution service at IRLEARN, send the command `SUBSCRIBE BIO-NAUT` followed by your full name as one line in the body of a mail message to `LISTSERV@IRLEARN.BITNET`. Users who are directly on EARN or BITNET can subscribe with an interactive message. For example, on an IBM running VM-CMS, you would type in `TELL LISTSERV AT IRLEARN SUB BIO-NAUT` plus your full name; on a VAX running VMS, you would enter `SEND LISTSERV@IRLEARN SUB BIO-NAUT` plus your name.

After you have sent one of those messages, you receive an acknowledgment that you now belong to the distribution list. Any mail messages to `BIO-NAUT@IRLEARN.BITNET` (`BIO-NAUT%IRLEARN.BITNET@VM1.NODAK.EDU` if you are an Internet user) are automatically distributed to all the other subscribers. Replies sent to one of those addresses are also seen by all members. Notice that the address for subscribing is different from the address to which you send messages.

> Moderator: Rob Harper
> `<HARPER@CSC.FI>`

BIO-SOFTWARE%NET.BIO.NET@VM1.NODAK.EDU
bio-soft@IRLEARN.UCD.IE   Ireland   EARN/BitNet
bio-soft@UK.AC.DARESBURY   U.K.   JANET
bio-software@BMC.UU.SE   Sweden   Internet
bio-software@NET.BIO.NET   U.S.A.   BitNet
bio-soft@NET.BIO.NET   U.S.A.   BitNet

The BIO-SOFTWARE mailing list replaces three previous mailing lists on BIOSCI (see Table B-1).

The BIO-SOFTWARE mailing list is not moderated. Questions, answers, and discussion are welcomed about software related to the biological sciences. Even software used indirectly in the course of doing research, such as word processors and communications software, can be discussed on this list.

To subscribe, send a request to one of the addresses in Table B-2. If you have ever received e-mail from any of the three groups that BIO-SOFTWARE replaced, you are probably already signed up with BIO-SOFTWARE. If you are not sure whether you are

TABLE B-1 Past and Present Names of BIO-SOFT-WARE Mailing Lists	Old BBS Name	BITNET/ EARN Name	Usenet Newsgroup Name
	CONTRIBUTED-SOFTWARE	SOFT-CON	bionet.software.contrib
	PC-COMMUNICATIONS	SOFT-COM	bionet.software.pc.comm
	PC-SOFTWARE	SOFT-PC	bionet.software.pc
	New BBS Name		
	BIO-SOFTWARE	BIO-SOFT	bionet.software

already a subscriber, contact the applicable BIOSCI address in Table B-2.

TABLE B-2 BIO-SOFT-WARE Subscription Addresses	Address	Location	Network
	biosci%net.bio.net@VM1.NODAK.ED		Internet
	biosci@net.bio.net	U.S.A.	BITNET
	biosci@irlearn.ucd.ie	Ireland	EARN/BITNET
	biosci@uk.ac.daresbury	U.K.	Janet
	biosci@bmc.uu.se	Sweden	Internet

Users on the Bionet DEC 2065 computer may subscribe by entering `DO BIO-SOFTWARE` after the @ prompt.

Coordinator: Dave Kristofferson
`<Kristofferson%BIONET-20.BIO.NET@VM1.NODAK.EDU>`
or
`<kristofferson@BIONET-20.BIO.NET>`

BIOMCH-L%HEARN.BITNET@CUNYVM.CUNY.EDU

This mailing list is for members of the International, European, American, Canadian and other Societies of Biomechanics and for other people with an interest in biomechanics and human and animal movement science for the scope of this list, see, e.g., the *Journal of Biomechanics* (Pergamon Press), the *Journal of Biomechanical Engineering* (ASME), and *Human Movement Science* (North-Holland).

To subscribe to BIOMCH-L, send the command `SUBSCRIBE BIOMCH-L` plus your full name for example, `SUBSCRIBE BIOMCH-L John Doe` to `LISTSERV%HEARN.BITNET@CUNYVM.CUNY.EDU` by e-mail or as interactive message.

> Coordinators:
> Anton J. van den Bogert
> `<WWDONIC%HEITUE5.BITNET@CUNYVM.CUNY.EDU>`
> Herman J. Woltring
> `<WWTMHJW%HEITUE5.BITNET@CUNYVM.CUNY.EDU>`

BIOSPH-L%UBVM.BITNET@VM1.NODAK.EDU

This list replaces the now defunct OZONE@ICNUCEVM. The new name reflects better the topics discussed. Anything relating to pollution, the CO_2 effect, ecology, habitats, climate, and related topics can be discussed.

To subscribe send the following command `SUBSCRIBE BIOSPH-L` plus your full name to `LISTSERV@UBVM.BITNET`. To leave the list, send `SIGNOFF BIOSPH-L`. If you are not a BITNET user, send commands to subscribe or cancel your subscription to `LISTSERV%UBVMS.BITNET@VM1.NODAK.EDU` in the body text of a message.

> Coordinator: Dave Phillips
> `<V184GAVW%UBVMS.BITNET@VM1.NODAK.EDU>`

BIOTECH%UMDC.BITNET@CUNYVM.CUNY.EDU
BIOTECH@UMDC.UMD.EDU

This Biotechnology mailing list is open for discussion of software and hardware issues, announcements, submission of bulletins, and exchange of ideas and data. Previous bulletins are archived on a BIOSERVE server disk; the server accepts commands from the Subject: line of a message. Requests for information and previous bulletins can be sent to `BIOSERVE%UMDC.BITNET@CUNYVM.CUNY.EDU` or `BIOSERVE@UMDC.UMD.EDU`.

All problems, questions, and requests to be added to or deleted from this list should be sent to `BIOTECH%UMDC.BITNET@CUNYVM.CUNY.EDU` or `BIOTECH@UMDC.UMD.EDU`.

> Coordinator: Deba Patnaik
> `<DEBA%UMDC.BITNET@CUNYVM.CUNY.EDU>`

BNFNET
`<Eng-Leong_Foo_%KOM.KOMunity.SE@VM1.NODAK.EDU>`
To join this discussion group on biological nitrogen fixation, send your full name, your postal address, and a short description of your work and your interest in BNF to the coordinator.

> Coordinator: Eng-leong Foo
> `<Eng-Leong_Foo_%KOM.KOMunity.SE@VM1.NODAK.EDU>`

EBCBBUL%HDETUD1.BITNET@VM1.NODAK.EDU
This bulletin board is run by the European Bank of Computer Programs in Biotechnology (EBCB), a nonprofit organization funded mainly by the European Community (EC). The main goal of EBCB is to stimulate and facilitate the use of computers in biotechnological training and research in Europe.

Items accepted for the bulletin board are distributed via e-mail to all participants and are also retained for future reference. EBCBBUL is public, and anyone with access to EARN can participate. To gain access to EBCBBUL through EARN or a related system, you issue one of the following commands: On EARN nodes operating with VM/CMS systems, type `TELL LISTSERV AT HDETUD1 SUBSCRIBE EBCBBUL` followed by your name.

On EARN nodes operating with VAX/VMS systems, type `SEND LISTSERV @ HDETUD1 SUBSCRIBE EBCBBUL` followed by your name. On other systems send mail to `LISTSERV@`

HDETUD1 or `LISTSERV@HDETUD1.TUDELFT.NL` (`LISTSERV%HDETUD1.BITNET@VM1.NODAK.EDU` if you are an Internet user). Include the command `SUB EBCBBUL` followed by your first and last names in the text of the mail.

As soon as your request for access has been accepted, you receive confirmation via e-mail. After that, you can use LDBASE or LSVTALK to search EBCBBUL (those two programs, available within the EARN node, make interactive searching possible).

> Coordinator: Arie Braat
> `<RCSTBRA%HDETUD1.TUDELFT.NL@VM1.NODAK.EDU>`
> or
> `<RCSTBRA@HDETUD1.TUDELFT.NL>`
> or
> `<EBCBBUL%HDETUD1.BITNET>`

ENVBEH-L@GRAF.POLY.EDU
ENVBEH-L%POLYGRAF.BITNET@MITVMA.MIT.EDU
ENVBEH-L@POLYGRAF (BITNET)

This mailing list is on environmental behavior: environment, design, and human behavior. Topics of discussion concern the relationships between people and their physical environments, including architecture and interior design, environmental stress (for example pollution and catastrophe), and human response to built and natural settings.

If you are a BITNET subscriber, you can join by sending the Listserv SUB command along with your name to the appropriate address—for example, `SEND LISTSERV@POLYGRAF SUB ENVBEH-L Jon Doe` or `TELL LISTSERV AT POLYGRAF SUB ENVBEH-L John Doe`. To be removed from the list, enter `SEND LISTSERV@POLYGRAF SIGNOFF ENVBEH-L` or `TELL LISTSERV AT POLYGRAF SIGNOFF ENVBEH-L`.

Those people without interactive access can send the Listserv command portion of the preceding lines as the first text line of a message. For example, `SUB ENVBEH-L Jane Doe` would be the only line in the body of a message to `LISTSERV%POLYGRAF`.

BITNET@MITVMA.MIT.EDU. As a last resort, send mail to one of the coordinators.

> Coordinators:
> Richard Wener
> `<????%POLYGRAF.BITNET@MITVMA.MIT.EDU>`
> Tony Monteiro
> `<MONTEIRO%POLYGRAF.BITNET@MITVMA.MIT.EDU>`

EPID-L%QUCDN.BITNET@CORNELLC.CCS.CORNELL.EDU

This mailing list is for the discussion of current topics in epidemiology and biostatistics.

If you are a BITNET user you can subscribe by sending the command SUBSCRIBE EPID-L followed by your full name to LISTSERV@QUCDN.BITNET. If you are not a BITNET user send the same command in the body of mail to LISTSERV%QUCDN.BITNET@CORNELLC.CCS.CORNELL.EDU. To be removed from the list, send the command UNSUBSCRIBE EPID-L. A list of archived files can be obtained by sending the command INDEX EPID-L.

> Coordinator: Robert C. James
> `<JAMESRC%QUCDN.BITNET@CORNELLC.CCS.CORNELL.EDU>`

ETHOLOGY%FINHUTC.BITNET@CUNYVM.CUNY.EDU

An unmoderated mailing list for the discussion of animal behavior and behavioral ecology. Possible topics include new or controversial theories, new research methods, and equipment. Announcements of books, papers, conferences, and new software for behavioral analysis are also encouraged.

BITNET, EARN, and NetNorth users can subscribe by sending the interactive message TELL LISTSERV AT FINHUTC SUB ETHOLOGY John Doe or SEND LISTSERV@FINHUTC SUB ETHOLOGY Jane Doe. If you are not at a BITNET site, or you cannot send interactive messages, send mail to `LISTSERV%FINHUTC.BITNET@CUNYVM.CUNY.EDU`; the first line that is not part of the header should read `SUBSCRIBE ETHOLOGY` followed by your full name.

> Coordinator: Jarmo Saarikko
> `<SAARIKKO%FINHUTC.BITNET@CUNYVM.CUNY.EDU>`

EVOLUTION@KESTREL.ARPA

This direct-distribution mailing list enables members to discuss the consequences of the theory of evolution, with an emphasis on mathematics, computer simulations, and literature. While creationists in this group are few in number, the list does not limit discussion of controversial topics. The coordinator assumes no responsibility for opinions expressed in the group. Archives are available only upon request to `EVOLUTION-REQUEST@KESTREL.ARPA`. All problems, questions, and requests to be added or deleted to this list should be sent to `EVOLUTION REQUEST@KESTREL.ARPA`.

> Coordinator:
> `king@KESTREL.ARPA`

POPULATION-BIOLOGY
\<biosci%NET.BIO.NET@VM1.NODAK.EDU>

Population Biology members discuss a synthesis of population ecology and population genetics, pursuing a unified theory to explain the structure, functioning and evolution of populations. Topics include ecology, population genetics, systematics, evolution, morphometry, interspecific competition, sociobiology, mathematical modeling, population regulation, pest control, speciation, chromosomal evolution, social behavior, statistical methodology, management of endangered species, applications of molecular biology techniques, and whatever topic you want to bring up. The type of species discussed is not limited and can include viruses, protokaryotes, plants, animals (including humans), mythic and extinct species, and computer-simulated species. Technical problems, book reviews, meeting announcements, and so on are also welcome.

The BITNET and Usenet names of POPULATION-BIOLOGY are POP-BIO and bionet.population-bio, respectively. If you wish to participate in the group, send your subscription request to the appropriate BIOSCI node (see Table B-3). More information on BIOSCI can also be requested at those addresses below:

TABLE B-3 POPULATION-BIOLOGY Subscription Addresses	Address	Location	Network
	biosci%NET.BIO.NET@VM1.NODAK.EDU		Internet
	biosci@net.bio.net	U.S.A.	BITNET
	biosci@irlearn.ucd.ie	Ireland	EARN/BITNET
	biosci@uk.ac.daresbury	U.K.	Janet
	biosci@bmc.uu.se	Sweden	EARN/Internet

A subscriptions request can also be sent to LISTSERV at IRLEARN by sending a message, note, or mail with the line `SUB POP$BIO` followed by your full name.

> Moderator: Vincent Bauchau
> <VINCENT%BUCLLN11.BITNET@VM1.NODAK.EDU>

FELINE-L
<UMNEWS%MAINE.BITNET@VM1.NODAK.EDU>

This mailing list facilitates discussion and dissemination of information about all members of the cat family. Emphasis is placed on large and feral cats—lions, tigers, leopards, and cougars—both in captivity and in the wild. However, discussion concerning any felines, including personal pets and household tabbies, is welcome. Appropriate topics include research papers dealing with cat populations; conservation issues; legislation; management in zoos, wildlife parks, and the like; care and feeding; veterinary considerations; research use of cats; animal rights concerns; showing, breeding, and training. The list is presently running on UMNEWS@MAINE. Its managers, however anticipate transferring it to their LISTSERV as soon as it is installed and available.

To post messages by means of e-mail, send them to `UMNEWS@MAINE` and include an entry either in the form `TO: FELINE-L Discussion UMNEWS@MAINE` or in the form `SUBJECT: UMBB.FELINE-L` followed by your subject entry. Non-BITNET users should use `UMNEWS%MAINE.BITNET@VM1.NODAK.EDU` instead of `UMNEWS@MAINE`.

To subscribe, send the message `TELL UMNEWS AT MAINE BBOARD SUBSCRIBE FELINE-L` followed by your full name. To leave the list, send the message `TELL UMNEWS AT MAINE BBOARD UNSUBSCRIBE FELINE-L`.

If you are going to be away from your computer for an extended period of time, the operators of this list ask that you please unsubscribe from the list and resubscribe when you return. To obtain a log of past discussions, send the following `TELL UMNEWS AT MAINE SENDME FELINE-L DIGEST FROM BBOARD ( ASIS )`. This retrieves the most recent 1000 lines. For earlier digests send `TELL UMNEWS AT MAINE SENDME FELINE-L VOLnnnnn FROM BBOARD ( ASIS )` in which nnnnn starts at 00001 for the oldest digest. Non-BITNET users should send requests to subscribe directly to the coordinator until such time as the LISTSERV is operational.

INFO-GCG%UTORONTO.BITNET@VM1.NODAK.EDU

This mailing list covers computer-aided molecular biology and is of particular interest to users and managers of the Genetics Computer Group software from the University of Wisconsin.

If you are a BITNET user you can subscribe by sending the command followed by your full name to `LISTSERV@UTORONTO SUBscribe INFO-GCG`. Non-BITNET users can subscribe by sending the above command in the body of a message to `LISTSERV%UTORONTO.BITNET@VM1.NODAK.EDU`.

> Coordinator: John Cargill
> `<CARGILL%UTOROCI.BITNET@VM1.NODAK.EDU>`
> or
> `<SYSJOHN%UTOROCI.BITNET@VM1.NODAK.EDU>`

Energy

ENERGY-L
<JO%ILNCRD.BITNET@CUNYVM.CUNY.EDU>

This BITNET newsletter covers energy research in Israel. To subscribe, send a message to `LISTSERV%TAUNIVM.BITNET@CUNYVM.CUNY.EDU` and include in the body of the letter the command `SUB ENERGY-L` followed by your title, first name and last name.

> Coordinator: Joseph van Zwaren de Zwarenstein
> `<JO%ILNCRD.BITNET@CUNYVM.CUNY.EDU>`

Physics

PHYS-L%UCF1VM.BITNET@CUNYVM.CUNY.EDU

This forum is for teachers of college and university physics courses. Topics of particular interest include using BITNET itself as the primary medium for delivery of university courses, innovative teaching laboratory experiments, and the use of microcomputers in the physics classroom. The list is open for public subscription.

To subscribe, send the command followed by your full name by interactive message to `LISTSERV@UCF1VM SUBSCRIBE PHYS-L`. You can also send the same command as the only text line of a mail message to `LISTSERV@UCF1VM`. No list-specific Internet contact address exists, but if necessary, mail can be sent to JIM%UCF1VM.BITNET@CUNYVM.CUNY.EDU for subscriptions or help.

> Coordinator: Dick Smith
> <FAC0069%UWF.BITNET@CUNYVM.CUNY.EDU>

PHYSIC-L
<JO%ILNCRD.BITNET@CUNYVM.CUNY.EDU>
This BITNET newsletter announces upcoming weekly colloquia and seminars in physics at all Israeli Universities except the Technion. Planned Israeli workshops and conferences in physics are also announced.

To subscribe, send the command `SUB PHYSIC-L` followed by your title, first name, and last name in the body of a letter to `LISTSERV%TAUNIVM.BITNET@CUNYVM.CUNY.EDU`.

> Coordinator: Joseph van Zwaren de Zwarenstein
> <JO%ILNCRD.BITNET@CUNYVM.CUNY.EDU>

PHYSICS@SRI-UNIX.ARPA (or @MC.LCS.MIT.EDU)
This group is for the discussion of physics, with reasonable speculation allowed.

Archives are maintained on MIT-MC in the following files:

COMAIL;PHYS FILE   (current messages)

COMAIL;PHYS FILE00 (oldest archives)

COMAIL;PHYS FILE*nn* (next-oldest archives)

COMAIL;PHYS FILE06 (newest archives)

All problems, questions, and requests to be added to or deleted from this list should be sent to `PHYSICS-REQUEST@SRI-UNIX.ARPA` or `@MC.LCS.MIT.EDU`.

> Coordinator: Andrew Knutsen
> <knutsen@SRI-UNIX.ARPA>

POLYMERP%RUTVM1.BITNET@CUNYVM.CUNY.EDU

A discussion on Polymer Physics. Topics include meetings, articles, software, theories, materials, methods, tools, and polymer properties such as solubility, viscosity, self-diffusion, and absorption.

All requests to be added to or deleted from this list should be sent to the automatic server `LISTSERV%RUTVM1.BITNET@CUNYVM.CUNY.EDU` as commands (one per line) in the mail body or note. Valid commands are `INFO`, `HELP`, `LIST`, `SUBSCRIBE POLYMERP` followed by your full name, `UNSUBSCRIBE POLYMERP`, `REVIEW POLYMERP`, and `GET filename filetype`. Archives and data are kept on the same server (send `GET POLYMERP FILELIST` for a list of files).

> Coordinator: Jan Scheutjens
> `<SCHEUTJE%HWALHW50.BITNET@CUNYVM.CUNY.EDU>`

SovNet-L%INDYCMS.BITNET@CUNYVM.CUNY.EDU

SovNet-L is a public discussion and distribution list dedicated to the dissemination and exchange of nonclassified information on electronic communication to, from, and within the USSR. The list covers all forms of electronic communication and includes discussions on Soviet electronic mail begun on RusTeX-L. SovNet-L is presently unedited and unmoderated.

To subscribe, send a request to `ListServ@IndyCMS.BITNET` (Internet users to `LISTSERV%INDYCMS.BITNET@CUNYVM.CUNY.EDU`) including the following message text: `Sub SovNet-L` plus your full name.

> List Coordinator: John B. Harlan
> `<JBHarlan@IUBACS>`

fusion@ZORCH.SF-BAY.ORG
fusion%zorch@AMES.ARC.NASA.GOV
fusion%zorch.sf-bay.org@RELAY.CS.NET

This list is for the discussion of nuclear fusion. Contributions sent to the list are automatically archived. BITNET users can obtain a list of available archive files by sending the command `INDEX FUSION` to `LISTSERV@NDSUVM1`. The files can be retrieved by sending the commmand `GET FUSION` *filetype* or by using the database search facilities of LISTSERV. Send the command `INFO DATABASE` for more information on the latter.

You can also obtain copies of the list's notebooks via anonymous ftp to `VM1.NODAK.EDU` (192.33.18.30); enter *anonymous* as the user ID, and use any password. Once you are validated enter `CD LISTARCH` and `DIR FUSION.*` to see the notebooks available. The file system is not hierarchical, so you must enter `CD ANONYMOUS` to return to the "root."

If you are a BITNET user you may subscribe by sending the command `SUB FUSION` followed by your full name (for example, `SUB FUSION John Doe`) to `LISTSERV@NDSUVM1` (or `LISTSERV@VM1.NODAK.EDU`). The command may be in the body, or text, of the mail (it may not be the subject) or may be sent interactively (via `TELL` or `SEND`) to `LISTSERV` on **BITNET**, **EARN**, or **NetNorth**. Non-BITNET users can join the list by sending `fusion-request@ZORCH.SF-Bay.ORG`.

> Coordinator: Scott Hazen Mueller
> `<scott@ZORCH.SF-BAY.ORG>`
> or
> `<scott%zorch@AMES.ARC.NASA.GOV>`
> or
> `<scott%zorch.sf-bay.org@RELAY.CS.NET>`

Geography

GEOGRAPH%FINHUTC.BITNET@CUNYVM.CUNY.EDU
GEOGRAPH on LISTSERV@FINHUTC

This global mailing list on geography was opened recently at the node FINHUTC by a group of young university geographers in Helsinki, Finland. Subscriptions to this list are offered to anyone with skills or interests relevant to geography.

To subscribe, send the command `SUBscribe GEOGRAPH` followed by your full name (for example, `SUBscribe GEOGRAPH Jane Doe`) to LISTSERV@FINHUTC, via mail or interactive message (Internet users send mail only to, `LISTSERV%FINHUTC.BITNET@CUNYVM.CUNY.EDU`). It is important to send the subscription command to `LISTSERV@ FINHUTC` and not to `GEOGRAPH@FINHUTC`. Contributions to the discussion list, however, are sent to `GEOGRAPH@FINHUTC`.

More information about the list can be obtained from `PKOKKONE@1FINUHA` (Pellervo Kokkonen) or `PYYHTIA@FINUHA` (Mervi Pyyhtia).

If you have a technical question, please send the command `HELP` to `LISTSERV@FINHUTC` or to your nearest listserver. (For more specific information, use the command `HELP topic`).

Technology Issues

INFO-FUTURES@ENCORE.COM
...{bu-cs,decvax,necntc,talcott}!encore!info-futures (UUCP)

This digest provides a speculative forum for analyzing current and likely events in technology as they affect our near future in computing and related areas. In broad terms, topics of interest include developments in both research and industry that are likely to affect our decision making. All problems, questions, and requests to be added to or deleted from this list should be sent to `INFO-FUTURESREQUEST@ENCORE.COM`.

> Moderator: Barry Shein
> `<bzs@ENCORE.COM>`

Chemistry

ORGCHE-L

This mailing list on organic chemistry facilitates the interchange of ideas, information, computer programs, and papers, and announces opportunities for collaborative efforts (in teaching or research) among specialists in organic chemistry and related areas.

To subscribe, send mail including the line `SUBS ORGCHE-L` followed by your full name to `LISTSERV%RPICICGE.BITNET@CUNYVM.CUNY.EDU`. If you do not receive mail confirming your subscription, contact `MSMITH%AMHERST.BITNET@CUNYVM.CUNY.EDU`, and your name will be added to the list.

> Coordinator: Asuncion Valles
> `<D3QOAVC0%EB0UB011.BITNET@CUNYVM.CUNY.EDU>`

Environment

SEAC-L%UNCVX1.BITNET@CUNYVM.CUNY.EDU

This list is for members of local chapters of the Student Environmental Action Council, (SEAC) and for students interested in forming chapters of SEAC on their campuses. Topics include actions taken by local chapters, coordination of national efforts, conferences, and bulletins of scientific interest on environmental topics.

To subscribe, send the command `SUB SEAC-L` followed by your full name to `LISTSERV@UGA` (for example, `SUB SEAC-L John Doe`). To have your name removed from the SEAC-L subscriber list, send `SIGNOFF SEAC-L`. Commands can be sent to

LISTSERV@UGA either as interactive messages or as e-mail (one command per line in the body of the e-mail message). Subscription problems or questions may be directed to one of the list coordinators.

> Coordinators:
> UNC SEAC
> `<seac%UNC.BITNET@CUNYVM.CUNY.EDU>`
> Paul Jones
> `<pjones%UNCVX1.BITNET@CUNYVM.CUNY.EDU>`

SFER@MTHVAX.CS.MIAMI.EDU

The South Florida Environmental Reader is intended primarily to keep people in South Florida abreast of local environmental issues. The newsletter is published on a monthly basis, in both electronic and paper formats.

To receive the electronic edition, send a message to `sfer-request@MTHVAX.CS.MIAMI.EDU` or to `SFER@UMIAMI.BITNET`.

> Coordinator: A. E. Mossberg
> `<aem@MTHVAX.CS.MIAMI.EDU>`

Space Science

SPACE@ANDREW.CMU.EDU

A daily digest on space-related topics. Requests for back issues should be directed to `SPACEREQUEST@ANDREW.CMU.EDU`. All problems, questions, and requests to be added to or deleted from this list should be sent to `SPACEREQUEST@ANDREW.CMU.EDU`. A BITNET subdistribution list, SPACE@ UGA, can be joined by BITNET subscribers by sending the `SUB` command along with your name. (For example, `SEND LISTSERV@UGA SUB SPACE JANE DOE`). To be removed from the list, enter `SEND LISTSERV@UGA SIGNOFF`. To contribute to the list,

BITNET subscribers should send mail to the Internet list name, not to the BITNET list name.

> Coordinator: Owen T. (Ted) Anderson
> `<OTA@ANDREW.CMU.EDU>`

APPENDIX C

Gateway Services to the Networks

For a subscription fee, gateway services provide access to networks such as the Internet, BITNET, and Usenet. If you have no other means to access those networks, such as affiliation with an academic or research institution, a gateway service can help.

The most valuable resource that gateway services provide is the ability to send electronic mail—through gateways—to millions of individuals on networks around the world.

DASnet

DASnet provides a mail and file transfer gateway to more than 60 commercial e-mail services and networks that are not otherwise directly linked with each other. Each subscribing network allows its users to send mail and distribution lists through DASnet's gateway to any other subscribing network or service.

For example, I can write a letter on EcoNet and send it to a person on The WELL. DASnet translates the different mail formats from one network's to the other's. DASnet users can also receive mail and have it delivered by telex, fax, and even the U.S. Post Office. DASnet reaches about 4 million people.

DASnet currently has access to ABA/net, Alternex (Brazil), ATT Mail, BITNET, BIX, CESAC (Italy), CIGnet, ComNet (Switzerland), CONNECT, DASnet Network, Deutsche Mailbox (Germany), Dialcom, FredsNaetet (Sweden), GeoNet (Germany, United Kingdom, United States), GreenNet (United Kingdom), INET, INFOTAP (Luxembourg), Mailbox Benelux (Holland), MBK Mediabox (Germany), MCI Mail, Nicarao (Nicaragua), NWI, PeaceNet/EcoNet, Pegasus (Australia), PINET, Portal, PsychNet, MetaNet, San Francisco/Moscow Teleport (U.S.S.R./United States), Telemail network, Telephone (France), TEXTEL (Caribbean), TWICS (Japan), UNISON, UUCP, Web (Canada), The WELL, Internet Domains, and networks whose internal e-mail systems are linked to DASnet services.

How to Get Online
If your network is not a DASnet subscriber, you can subscribe yourself. Contact DA Systems, Inc., 1503 E. Campbell Ave., Campbell, CA 95008; 408/559-7434.

Cost
The initial start-up fee is $33.50, which includes a $20 deposit and a DASnet user directory. After that, you pay $4.75 a month, plus usage (message) time. Message charges depend on the network, but the average is 45 cents for the first 1000 characters (about one single-spaced page) and 20 cents for each additional 1000. Overseas costs are higher. File transfers cost $15.75 each.

UUNET

UUNET Technologies is a nonprofit organization that provides access for a fee to the Internet, UUCP mail, Usenet, and affiliated network services. UUNET provides some 700 news feeds and 1600 direct mail connections. If you have your own business or computer and want to become an established domain, UUNET can register you (for $35) and act as your Internet mail forwarder.

AlterNet is UUNET's approach to bringing the Internet to the public. The Internet is primarily funded by the U.S. Government, and the network's use is restricted to organizations doing government-sponsored or approved research and development. AlterNet is a cost-effective way for an organization to become part of the Internet.

UUNET also sells network-related publications and Telebit Trailblazer modems at a substantial discount.

How to Get Online
For subscription information, contact UUNET Communications Services, 3110 Fairview Park Dr., Ste. 570, Falls Church, VA 22042; 703/876-5050.

Cost
You can connect to UUNET as a local call from thousands of American cities via the CompuServe telecommunications network, or you can connect directly. A toll-free number is available for $10 per hour for nighttime use. Otherwise, you are billed a $35 monthly fee plus $2 per hour connect time if you call them directly.

Even if you do not subscribe to UUNET, you can call its 900 number, for 40 cents a minute, to log on and download more than 600 megabytes' worth of UNIX source archives. The sources are also available for purchase on tape.

UUNET offers several pricing schemes, including a $20 per month subscription for low-volume users that includes two hours of connect time per month.

DIAL n' CERF

CERFnet (California Education and Research Federation Network), a midlevel network of NSFNET that links more than 70 research and education centers, has initiated a service called DIAL n' CERF. This dial-up service provides Internet access to organizations and individuals who have legitimate needs, want short-term access, or already have access but need access when traveling. Users have access to terminal services (telnet), SLIP (Serial Line Internet Protocol, a process that allows you to use a modem to connect to the Internet host), Usenet news, and Internet mailboxes.

How to Get Online
Contact CERFnet, San Diego Supercomputer Center, P.O. Box 85608, San Diego, CA 92186; or call 619/534-5087 (or 800/876-CERF). You can reach them by electronic mail at help@cerf.net.

Cost
DIAL n' CERF is an affordable way to obtain Internet access. Costs to connect to one of the five CERFnet locations in California include a one-time installation fee per organization of $250 (regardless of how many users are on the system), a monthly fee of $20 per user, and connect charges of $5 per hour.

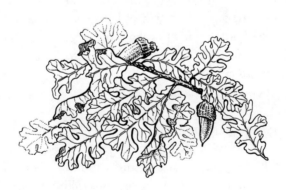

APPENDIX D

Sample BBS Session

The following is a complete transcript of an eight-minute BBS session. The bulletin board accessed is the FDA/CDRH Guidelines BBS, sponsored by the Food and Drug Administration. User input is boldfaced.

This transcript is presented to give an idea of what to expect during a bulletin board session. Remember, every BBS is different, reflecting the system itself and the interests, creativity, and idiosyncracies of its SysOp.

```
CONNECT 2400
WELCOME TO FDA/CDRH GUIDELINES



What is your FIRST Name? don
What is your LAST Name? rittner
Checking Users...
What is your CITY and STATE? schenectady, ny
Welcome to the world of RBBS-PC!  However, before continuing
you should understand your responsibilities as a RBBS-PC
user.  Specifically they are:
```

1. Actively encourage and promote the free exchange and discussion of information, ideas and opinions, except when the content would compromise the national security of the United States; violate proprietary rights, personal privacy, or applicable state/federal/local laws and regulations affecting telecommunications; or constitute a crime or libel.

2. Use your real name and fully disclose any personal, financial, or commercial interest when evaluating any specific product or service.

3. Adhere to these rules and notify me immediately when you discover any violations of the rules.
FURTHER every user explicitly acknowledges that all information obtained from this RBBS-PC is provided "as is" without warranty of any kind, either expressed or implied, including, but not limited to the implied warranties of merchantability and fitness for a particular purpose and that the entire risk of acting on information obtained from this RBBS-PC, including the entire costs of all necessary remedies, is with those who choose to act on such information and not the operator of this RBBS-PC.
Press [ENTER] to continue?

DON RITTNER from SCHENECTADY, NY
<C>hange name/address, <D>isconnect, <R>egister? **r**
Enter PASSWORD you'll use to logon again? **test**
Re-enter PASSWORD for verification (Dots Echo)? ....

Please REMEMBER your password
CAN YOUR TERMINAL DISPLAY LOWER CASE (Y/N)? **y**
UPPER CASE and lower

   * <Ctrl K>/<Ctrl X> aborts <Ctrl S> suspends *
   ================ GRAPHICS HELP FILE ====================

Three different configurations can be selected for you to use for your menus while using RBBS-PC. We list them as "NONE", "ASCII" & "COLOR".

What does that mean to you? Well the "NONE" menus can use all the characters in the alphabet plus numbers. The "ASCII" menus contain all the characters supported by IBM/PC computers. The "COLOR" menus contain all of the above plus escape sequences that (with the proper communications program) will present you with color menus.

All menus contain the same options, so it is not important which one you select. It's just our way of showing off our systems.

It should be noted that the communication program "YOU" use makes the difference on what you see on your end of the line. I have included
MORE: [Y],N,NS? **Y**

a table to let you know ahead of time what to expect.

	TEXT	ASCII	COLOR
Smartcom	Y	N	N
Xtalk	Y	N	N
Execpct	Y	Y	Y
Pctk666	Y	Y	Y
Qmodem	Y	Y	Y

GRAPHICS wanted: N)one, A)scii-IBM, C)olor-IBM, H)elp? **n**
GRAPHICS: None

Default Protocol:
A)scii, X)modem, C)Xmodem/CRC,
K)ermit,N)one? **x**
PROTOCOL: Xmodem

```
Want nulls (for printing terminal) (Y/N)? n
Nulls Off
Logging DON RITTNER
RBBS-PC CPC15.1B NODE 1
OPERATING AT 2400 BAUD,N,8,1


* <Ctrl K>/<Ctrl X> aborts <Ctrl S> suspends *
+-----------------------------------------------+
|  Welcome to the world of RBBS-PC — Dedicated to the
|  free exchange of information.  Your SYSOP is FDA's
|  Division of Mechanics & Materials Science. (Ed
|  Mueller)
|  Voice: 301-443-7003        Data: 301-443-7496
|
+-----------------------------------------------+
```

This bulletin board is based on an IBM PC and contains draft FDA guidance documents which are available for downloading. If you find a problem with the bulletin board, please leave a message using the "C"omments command or call the FDA on 301-443-7003.

This system answers the telephone at 300/1200/2400 baud, no parity, eight data bits, and 1 stop bit.
Files Downloaded: 0 Uploaded: 0

8 NEW BULLETIN(S) since last call: 1 2 3 4 5 6 7 8
READ new bulletins (Y=[ENTER],N)?

* <Ctrl K>/<Ctrl X> aborts <Ctrl S> suspends *

The use of this PC bulletin board is being evaluated by the FDA for making available PMA, IDE, 510k and other guidance documents and to give interested members of the medical, scientific and industrial communities a direct method for providing FDA with suggested improvements.

These documents are stored on this bulletin board using either ASCII or WORDPERFECT 5.1 format.

Diagrams and figures referenced in the guidance files are incorporated into the guidance documents as PCX files. These files can be viewed and printed by WORDPERFECT 5.1.
MORE: [Y],N,NS? **Y**

To see which files are available on the bulletin board, you must first access the files subsystem by entering an "F" from the main menu. To list the available files enter an "L" and answer the question with an "A". You will be provided with a list of currently available FDA guidance files on the BBS.

To download a file enter a "D" and type the file name including the file extension that you wish to download.

Then set up your PC to receive the file using your file emulation software (PROCOMM, CROSSTALK, PCTALK etc.). If you wish to leave comments for the FDA, you can enter the "C" command from the main menu and type in your comments. To terminate your comments, enter two (2) successive "CR"s . Be sure to use the "S" save command to insure your comments are forwarded. If your comments are extensive, please incorporate them into a file and upload them to the bulletin board.

When uploading files, please leave a message for the FDA system operator explaining what your file pertains to and in which file format it is stored. DCA, WORDPERFECT 5.1 or ASCII formats are preferred.

If you have problems with the system or suggestions for improvement, please leave a message on the BBS or telephone the FDA Division of Mechanics and Materials Science on (301)-443-7003.

43 min left

* <Ctrl K>/<Ctrl X> aborts <Ctrl S> suspends *

I just obtained a copy of an ascii file which describes the use of the RBBS Bulletin Board in simple terms. The file name is USRGUIDE.BBS and is available for downloading from this BBS.

43 min left

* <Ctrl K>/<Ctrl X> aborts <Ctrl S> suspends *

A listing of CDRH contracts and reports available from NTIS is provided in an ASCII file named BBNTIS.ASC.
43 min left

* <Ctrl K>/<Ctrl X> aborts <Ctrl S> suspends *
Research Agenda for the 1990s Available

The Center for Devices and Radiological Health has drafted a compilation of potential research efforts which could fill significant gaps in scientific and technical knowledge about medical devices and radiation-emitting products. The Center is releasing this document, "CDRH Research Agenda for the 1990s," in draft form to solicit comments from and stimulate communication with the medical device and radiological health communities.

The Center's compliance and product review activities are often impeded by a lack of knowledge or data that would scientifically support regulatory decisions. Sometimes these decisions must be made in the face of considerable scientific uncertainty; other times, decisions must be delayed until the appropriate data can be developed. The gaps in knowledge range from basic biochemical, biophysical or statistical questions to practical methods for gauging device performance, safety, and effectiveness. The Center's Office of Science and Technology is able to fill some of these gaps through in-house research and by funding research outside the Center. Still, resources are limited and many gaps in knowledge are only partially resolved or are not addressed at all.

Therefore, Center staff have tried to identify the most important current and future research needs associated with medical devices and radiation-emitting products from an FDA perspective. The Agenda consists of descriptions of 38 potential research topics that the Center does not plan to address because of insufficient resources or expertise. The projects are divided into four broad categories:

* New concepts and data that would have immediate beneficial effects on the safety or effectiveness of medical devices. An example would be the development of physiological models and associated data bases suitable for developing and testing "intelligent" medical devices.

* Quantitative tools and test methods to assess the performance of existing medical devices or the safety of new ones. An example is the establishment of a national hydrophone calibration facility through which reproducible ultrasound exposure levels might be traced.

* Integration of data and information to address the growing needs for extensive data bases that provide reference for the assessment of older devices and the development of new ones. An example would be the establishment of a national retrieval and analysis system for explanted medical devices with the associated development of uniform protocols for testing, analysis and data inclusions.

* Addressing basic unanswered questions whose answers would not immediately impact on device safety and effectiveness but are ultimately necessary for improving existing devices or developing new ones. An example of such a question would be, "What are the chemical and physical mechanisms for thrombus formation near prosthetic heart valves and other cardiovascular devices?" The answer could be useful in the design of new heart valves, vascular prostheses, etc.

The Center is asking the medical device and radiological health communities to review the Agenda and comment on:

* Important research topics that should be added, and why.

* Research topics now on the list that really do not represent important gaps.

* Those knowledge gaps that may already have been filled.

In addition to stimulating a dialogue that will help round out the Center's knowledge base, the Research Agenda is also expected to guide researchers, and the manufacturers and public agencies who fund research, toward activities that will accomplish the most in terms of technical innovation, transfer of technology to practical application, and public health impact.

A copy of this document is available for downloading from this BBS under the name RESAGEND.WP5. The format for the file is Wordperfect 5.

For more information contact Harvey Rudolph, Office of Science and Technology (HFZ-140), CDRH, 5600 Fishers Lane, Rockville, MD 20857, telephone 301-443-3314.

```
43 min left
* <Ctrl K>/<Ctrl X> aborts <Ctrl S> suspends *
```

We are going to start using a file compression scheme for reducing the size of download files. This should save you telephone line charges and will be particularly important with guidance documents which contain figures. We
MORE: [Y],N,NS? **Y**

will be using Wordperfect 5.1 for all future documents and as time permits will be converting existing documents over to 5.1 format. Use of this format allows us to incorporate both figures and equations easily into the documents. The compression format we will be using is the ZIP file compression format. A program called PKUNZIP.EXE is available on this Bulletin Board for decompressing the file. To use it make sure the program is in the same directory as the file you wish to decompress and type

PKUNZIP filename.zip

By way of example the only file on this BBS today in ZIP format is the Latest version of the Heart Valve Guidance. This includes the figure files. To decompress it type

PKUNZIP HVX.ZIP

The HVX.ZIP will decompress to 5 files called HVW2.w51, TEST1.PCX, TEST2.PCX TEST3.PCX and INFO.W51. The HVW2.w51 is the Wordperfect 5.1 guidance document and the other PCX files are the figures referenced in the document.The INFO.W51 is a single page which provides a contact point in FDA for questions concerning the guidance.

If you have any questions about these changes please contact the sys-op on the BBS by leaving a message or calling me (Ed Mueller) 301-443-7003.

```
   43 min left
 * <Ctrl K>/<Ctrl X> aborts <Ctrl S> suspends *


        CENTER FOR DEVICES AND RADIOLOGICAL HEALTH
               ADVISORY COMMMITTEE MEETINGS


 JULY 1990

 July 11            DENTAL PRODUCTS PANEL
                    July 11, 8 a.m. - 4 p.m.
                    Parklawn Building, Conference Room E
                    5600 Fishers Lane
                    Rockville, MD
 July 18            GASTROENTEROLOGY-UROLOGY DEVICES PANEL
                    July 18  9 a.m. - 4:30 p.m.
                    Piccard Building, First Floor
                    Conference Room, Rockville, MD


 AUGUST 1990

 August 20 & 21     CIRCULATORY SYSTEM DEVICES PANEL
                    August 20 and 21, 8:30am - 4:30pm
                    (both days) Hubert Humphrey Building,
                    Rooms 503-529A
                    200 Independence Avenue, SW
                    Washington, DC


   42 min left
 * <Ctrl K>/<Ctrl X> aborts <Ctrl S> suspends *
```

The FDA Medical devices Standards Activities Report is now available on this BBS. The document is compressed using the ZIP file compression program. After being uncompressed using the PKUNZIP program which is available for downloading from this BBS the file will be in Wordperfect 5.1 format. This standards file is a comprehensive listing of current national and international, voluntary and regulatory standards activities in the field of medical devices.

```
42 min left
* <Ctrl K>/<Ctrl X> aborts <Ctrl S> suspends *

MORE: [Y],N,NS? Y
```

NEW DOCUMENTS ADDED TO THIS BBS

This bulletin will be updated each month to indicate which new documents have been added to the BBS since the last update period.

JUNE 1990

1. STDS.ZIP   FDA MEDICAL DEVICE SURVEY OF NATIONAL & INTERNATIONAL, VOLUNTARY & REGULATORY STANDARDS ACTIVITIES.

2. VASULAR.W51 REVISED FORM OF FDA VASCULAR GRAFT PMA/510k/IDE GUIDANCE.

3. IAB.W51   GUIDANCE FOR DETERMINING THE EQUIVALENCE OF INTRAAORTIC BALLOON CATHETERS UNDER 510K.

4. HVX.ZIP   REVISED GUIDANCE FOR REPLACEMENT HEART VALVE PMA REVIEW.(INCLUDES NEW PROVISIONS FOR CLINICAL DOPPLER HEMODYNAMICS DATA).

5. PRETGF.W51   PREMARKET TESTING GUIDELINES FOR FEMALE BARRIER CONTRACEPTIVE DEVICES.

6. APR.W51   GUIDANCE FOR THE PREPARATION OF THE ANNUAL REPORT TO THE PMA APPROVED HEART VALVE PROSTHESES

7. IMAG.W51   RADIATION EXPOSURE FROM MEDICAL IMAGING DEMONSTRATIONS

8. CFR.W51   EXEMPTION FROM REPORTING UNDER 21 CFR 1002

9. TRANS.W51   TRANSDUCER LABELING REQUIREMENTS FOR DOPPLER ULTRASOUND EQUIPMENT

10. BIOMAT.W51 BIOMATERIALS WORKSHOP PRINCETON REGIONAL MEETING MARCH 22, 1990

```
42 min left
```

```
Bulletin #(s) [1 thru 10], L)ist, N)ew (Press [ENTER] to
quit)?
Checking messages in MAIN.
Sorry, DONALD, NO MAIL for you


RBBS-PC CPC15.1B Node   1


Caller #   1223    # active msgs: 23   Next msg # 78
42 min left

                 RBBS-PC   MESSAGE   SYSTEM
           ~~~~~~~~~~~~~~~~~~~~~~~~~~~~~~~~~~~
        --- COMMUNICATIONS -- - UTILITIES - - ELSEWHERE -
        PERSONAL MAIL       SYSTEM COMMANDS
        E)nter a Message    A)nswer Questions  H)elp        D)oors
        K)ill a Message     B)ulletins         J)oin        Subsystem
        P)ersonal Mail      C)omment           Conferences  F)iles
           Found            I)nitial Welcome   V)iew        Subsystem
        R)ead Messages      O)perator Page     Conferences  G)oodbye
        S)can Messages      W)ho else is on    X)pert on/off Q)uit to
        T)opic of Msgs                         ?)List       other Sub-
           shown                                  Functions systems
                                                            U)tilities
                                                            Subsystem


MAIN command <?,A,B,C,D,E,F,H,I,J,K,O,P,Q,R,S,T,U,V,W,X>? b
* <Ctrl K>/<Ctrl X> aborts <Ctrl S> suspends *
================== Bulletin Menu ==================
         1 - DESCRIPTION OF BULLETIN BOARD.
         2 - BBS USER GUIDE
         3 - CDRH NTIS PUBLICATIONS LIST
         4 - CDRH RESEARCH AGENDA FOR THE 1990s
         5 - FILE COMPRESSION/ IMAGE FILES
         6 - CDRH ADVISORY COMMITTEE MEETINGS
             SCHEDULE FOR JULY/AUGUST 1990
         7 - FDA MEDICAL DEVICES STANDARDS
```

ACTIVITIES REPORT
8 - NEW DOCUMENT ADDITIONS TO BBS

Bulletin #(s) [1 thru 10], L)ist, N)ew (Press [ENTER] to quit)? **1**

* <Ctrl K>/<Ctrl X> aborts <Ctrl S> suspends *

The use of this PC bulletin board is being evaluated by the FDA for making available PMA, IDE, 510k and other guidance documents and to give interested members of the medical, scientific and industrial communities a direct method for providing FDA with suggested improvements. These documents are stored on this bulletin board using either ASCII or WORD-PERFECT 5.1 format.

Diagrams and figures referenced in the guidance files are incorporated into the guidance documents as PCX files. These files can be viewed and printed by WORDPERFECT 5.1.

MORE: [Y],N,NS? **Y**

To see which files are available on the bulletin board, you must first access the files subsystem by entering an "F" from the main menu. To list the available files enter an "L" and answer the question with an "A". You will be provided with a list of currently available FDA guidance files on the BBS.

To download a file enter a "D" and type the file name including the file extension that you wish to download.

Then set up your PC to receive the file using your file emulation software (PROCOMM, CROSSTALK, PCTALK etc.). If you wish to leave comments for the FDA, you can enter the "C" command from the main menu and type in your comments. To terminate your comments, enter
MORE: [Y],N,NS? **Y**

two (2) successive "CR"s . Be sure to use the "S" save command to insure your comments are forwarded. If your comments are extensive, please incorporate them into a file and upload them to the bulletin board.

When uploading files, please leave a message for the FDA system operator explaining what your file pertains to and in which file format it is stored. DCA, WORDPERFECT 5.1 or ASCII formats are preferred.

If you have problems with the system or suggestions for improvement, please leave a message on the BBS or telephone the FDA Division of Mechanics and Materials Science on (301)-443-7003.

```
42 min left
Bulletin #(s) [1 thru 10], L)ist, N)ew (Press [ENTER] to
quit)? L

=================== Bulletin Menu ==================
         1 - DESCRIPTION OF BULLETIN BOARD.
         2 - BBS USER GUIDE
         3 - CDRH NTIS PUBLICATIONS LIST
         4 - CDRH RESEARCH AGENDA FOR THE 1990s
         5 - FILE COMPRESSION/ IMAGE FILES
         6 - CDRH ADVISORY COMMITTEE MEETINGS
             SCHEDULE FOR JULY/AUGUST 1990
         7 - FDA MEDICAL DEVICES STANDARDS
             ACTIVITIES REPORT
         8 - NEW DOCUMENT ADDITIONS TO BBS

Bulletin #(s) [1 thru 10], L)ist, N)ew (Press [ENTER] to
quit)? 3

* <Ctrl K>/<Ctrl X> aborts <Ctrl S> suspends *
```

A listing of CDRH contracts and reports available from NTIS is provided in an ASCII file named BBNTIS.ASC.

42 min left

Bulletin #(s) [1 thru 10], L)ist, N)ew (Press [ENTER] to quit)? **L**

=================== Bulletin Menu ================
```
        1 - DESCRIPTION OF BULLETIN BOARD.
        2 - BBS USER GUIDE
        3 - CDRH NTIS PUBLICATIONS LIST
        4 - CDRH RESEARCH AGENDA FOR THE 1990s
        5 - FILE COMPRESSION/ IMAGE FILES
        6 - CDRH ADVISORY COMMITTEE MEETINGS
            SCHEDULE FOR JULY/AUGUST 1990
        7 - FDA MEDICAL DEVICES STANDARDS
            ACTIVITIES REPORT
        8 - NEW DOCUMENT ADDITIONS TO BBS
```

Bulletin #(s) [1 thru 10], L)ist, N)ew (Press [ENTER] to quit)?

41 min left

```
              RBBS-PC   M E S S A G E   S Y S T E M
     ~~~~~~~~~~~~~~~~~~~~~~~~~~~~~~~~~~~~~~
--- C O M M U N I C A T I O N S -- - UTILITIES - - ELSEWHERE -
PERSONAL MAIL     SYSTEM COMMANDS
E)nter a Message  A)nswer Questions  H)elp        D)oors
K)ill a Message   B)ulletins         J)oin            Subsystem
P)ersonal Mail    C)omment              Conferences  F)iles
   Found          I)nitial Welcome   V)iew            Subsystem
R)ead Messages    O)perator Page        Conferences  G)oodbye
S)can Messages    W)ho else is on    X)pert on/off Q)uit to
T)opic of Msgs                       ?)List           other Sub-
   shown                                Functions     systems
                                                   U)tilities
                                                      Subsystem
```

```
MAIN command <?,A,B,C,D,E,F,H,I,J,K,O,P,Q,R,  S,T,U,V,W,X>? f

41 min left

                RBBS-PC   F I L E   S Y S T E M
             ~~~~~~~~~~~~~~~~~~~~~~~~~~~~~~~
  ----- F I L E   T R A N S F E R ---- - UTILITIES - ELSEWHERE -
  FILE TRANSFER     FILE INFORMATION
  D)ownload a file  L)ist files available  H)elp      G)oodbye
  U)pload a file    N)ew files listed      X)pert     Q)uit to
                    S)earch file             on/off     other sub-
                      directories                       systems
                    V)iew ARC Contents
                    ?)File transfer tutorial


FILE command <?,D,G,H,L,N,Q,S,U,V,X>? L

What directories (<A>ll,[ENTER] for menu)? A

* <Ctrl K>/<Ctrl X> aborts <Ctrl S> suspends *

         Directories Available for Downloading

  DIRECTORY         CONTENTS

  Guide1            This Directory contains FDA Guidance Docu-
                    ments for PMA, 510K and IDE submissions. To
                    see the contents of this directory Type
                    L;GUIDE1


40 min left
```

```
                    RBBS-PC  F I L E  S Y S T E M
                    ~~~~~~~~~~~~~~~~~~~~~~~~~~~~~~

----- F I L E   T R A N S F E R ---- - UTILITIES - ELSEWHERE -
FILE TRANSFER       FILE INFORMATION
D)ownload a file    L)ist files available  H)elp      G)oodbye
U)pload a file      N)ew files listed      X)pert     Q)uit to
                    S)earch file           on/off     other sub-
                      directories                     systems
                    V)iew ARC Contents
                    ?)File transfer tutorial


FILE command <?,D,G,H,L,N,Q,S,U,V,X>? n
Files on/after (MMDDYY, [ENTER] = last on )?
What directories (<A>ll,[ENTER] quits)? a


Scanning Directory GUIDE1 for

    HVX.ZIP        05-07-90  Draft FDA Guidance for  Artificial
    AH.ASC         02-24-88  Draft FDA Guidance for Artificial
    ULTRA.ASC      02-24-88  Draft FDA Guidance For Diagnostic
    COCH.ASC       02-24-88  Draft FDA Guidance for PMA Application
    OXY.ASC        06-06-88  Guidance for Applications for Blood
    PTCA.W51       06-06-88  PTCA Catheter System Testing Guidance -
    FETELC.ASC     06-06-88  Guidance for Evaluation of Fetal
    LAP.ASC        06-06-88  Guidelines for Evaluation of Laparoscope
    IUD.ASC        06-06-88  Guidelines for Evaluation of Non-Drug
                              IUDs -
    ULT.W51        06-22-88  Proposed Regulatory Approaches for
    HSD.ASC        07-11-88  Guidelines for Evaluation of
    ACOCH.ASC      07-11-88  Guidance for the Arrangement and
    TOD.ASC        07-26-88  Guidelines for the Evaluation of
    VASCULAR.W51   06-06-90  Outline for Vascular Graft PMA/IDE Guid-
                              ance
    LIGAMENT.ASC   11-01-88  Guidance Document for the Preparation of
    BGS.ASC        11-01-88  Guidance Document for the Preparation of
    510K.ASC       11-01-88  Guidance on The Center for Devices
```

CERAMXBA.W51 03-04-90 Guidance document for the preparation of
TRIPAR.WP5   10-06-89 Tripartitie Biocompatibility Guidance
PRETGF.W51   04-04-90 Premarket Testing Guidelines for
IAB.W51   05-25-90 Determining Equivalence of Intraaortic
APR.W51   04-30-90 Guidance for the Preparation of
IMAG.W51   04-16-90 Radiation Exposure From Medical
CFR.W51   02-24-86 Exemption From Reporting Under
TRANS.W51   10-30-87 Transducer Labeling Requirements for
BBNTIS.ASC   07-26-88 FDA NTIS Publications List ASCII FORMAT.
USRGUIDE.BBS 04-12-88 USER GUIDE for RBBS in ASCII FORMAT.
NEMA510K.ASC 04-21-88 510(k) format document for diagnostic
NEMA510K.WS2 04-21-88 510(k) format document for diagnostic
FED.ASC   10-06-89 Guidance for the Emergency Use of
                      Unapproved
PREPRO.WP5   03-13-90 Preproduction Quality Assurance Planning
SMPLFLK.WP5 10-06-89 Software Design Guidance
ANNRPT89.W51 02-28-90 Office of Science and Technology
HHS1.WPF   07-15-88 Reviewer Guidance For Computer-
GUIDE1.DIR   05-16-90 Listing of all guidance Available
STDS.ZIP   06-05-90 FDA Medical Devices Standards Activities
BIOMAT.W51   03-22-90 Biomaterials Workshop Princeton
Scanning Directory DIR for

39 min left

```
              RBBS-PC   FILE   SYSTEM
              ~~~~~~~~~~~~~~~~~~~~~~~~~~~~~~

----- F I L E    T R A N S F E R ---- - UTILITIES - ELSEWHERE -
FILE TRANSFER      FILE INFORMATION
D)ownload a file   L)ist files available H)elp      G)oodbye
U)pload a file     N)ew files listed     X)pert     Q)uit to
                   S)earch file          on/off     other sub-
                      directories                   systems
                   V)iew ARC Contents
                   ?)File transfer tutorial
```

FILE command <?,D,G,H,L,N,Q,S,U,V,X>? **q**

QUIT to F)ile, [M]ain, U)til section or S)ystem (hang up)
([ENTER]=M)? **u**

39 min left

         R B B S - P C   U T I L I T I E S   S Y S T E M
         ~~~~~~~~~~~~~~~~~~~~~~~~~~~~~~~~~~~~~~~

------------- U T I L I T I E S ------------ --- ELSEWHERE --
USER PROFILE/PREFERENCE   SYSTEM
F)ile transfer protocol   B)aud rate (300->450) D)oor
G)raphics   C)lock (time of day)   subsystem
L)ines per page   H)elp   F)ile
M)essage Margin   P)assword changes   subsystem
P)assword Changes   S)ystem statistics   Q)uit to
R)eview User's Preferences   U)ser log   other
T)oggle Options   subsystems
   Line feeds (on/off)
   Nulls (on/off)
   Prompt bell (on/off)
   Expert (on/off)

UTIL command <?,B,C,F,G,H,L,M,P,Q,R,S,T,U,X>? **d**
All doors locked!

38 min left

         R B B S - P C   U T I L I T I E S   S Y S T E M
         ~~~~~~~~~~~~~~~~~~~~~~~~~~~~~~~~~~~~~~~

------------- U T I L I T I E S ------------ --- ELSEWHERE --
USER PROFILE/PREFERENCE   SYSTEM
F)ile transfer protocol   B)aud rate (300->450) D)oor
G)raphics   C)lock (time of day)   subsystem
L)ines per page   H)elp   F)ile
M)essage Margin   P)assword changes   subsystem

```
P)assword Changes          S)ystem statistics    Q)uit to
R)eview User's Preferences U)ser log               other
T)oggle Options                                    subsystems
   Line feeds (on/off)
   Nulls (on/off)
   Prompt bell (on/off)
   Expert (on/off)
UTIL command <?,B,C,F,G,H,L,M,P,Q,R,S,T,U,X>? q
QUIT to F)ile, [M]ain, U)til section or S)ystem (hang up)
([ENTER]=M)? m


  38 min left

             RBBS-PC   MESSAGE   SYSTEM
      ~~~~~~~~~~~~~~~~~~~~~~~~~~~~~~~~~~~~
--- COMMUNICATIONS -- - UTILITIES - - ELSEWHERE -
PERSONAL MAIL      SYSTEM COMMANDS
E)nter a Message   A)nswer Questions  H)elp        D)oors
K)ill a Message    B)ulletins         J)oin          Subsystem
P)ersonal Mail     C)omment             Conferences F)iles
   Found           I)nitial Welcome   V)iew          Subsystem
R)ead Messages     O)perator Page       Conferences G)oodbye
S)can Messages     W)ho else is on    X)pert on/off Q)uit to
T)opic of Msgs                        ?)List          other Sub-
   shown                                Functions    systems
                                                    U)tilities
                                                      Subsystem


MAIN command <?,A,B,C,D,E,F,H,I,J,K,O,P,Q,R,S,T,U,V,W,X>? c

Leave a comment for EDWARD (Y/N)? y
Sending personal mail to SYSOP
```

```
Type comment 99 lines max (Press [ENTER] to quit)

[-----------------------------------------------------------------------]
 1: Howdy! I'm writing a book on using online resources for the scientific
 2: community and would like to add your board to the book. Do you have a
 3: history of the board, who uses it, what kind of files are here, purpose, etc.?
 4: Also do you mind if I use a sample log on for the book?  Thanks!
 5: don rittner
 6:

A)bort, C)ontinue, D)elete, E)dit, I)nsert, L)ist, M)argin, S)ave
Edit Sub-function <A,C,D,E,I,L,M,S,?>? s
Adding new msg # 78.....
37 min left
```

```
                     RBBS-PC   M E S S A G E   S Y S T E M
              ~~~~~~~~~~~~~~~~~~~~~~~~~~~~~~~~~~~~~~

--- C O M M U N I C A T I O N S -- - UTILITIES - - ELSEWHERE -
PERSONAL MAIL         SYSTEM COMMANDS
E)nter a Message      A)nswer Questions    H)elp           D)oors Subsystem
K)ill a Message       B)ulletins           J)oin           F)iles Subsystem
P)ersonal Mail        C)omment               Conferences   G)oodbye
   Found              I)nitial Welcome     V)iew           Q)uit to other Subsystems
R)ead Messages        O)perator Page         Conferences   U)tilities Subsystem
S)can Messages        W)ho else is on      X)pert on/off
T)opic of Msgs                              ?)List functions
   shown


MAIN command <?,A,B,C,D,E,F,H,I,J,K,O,P,Q,R,S,T,U,V,W,X>? g
Disconnect the call (Y,N=[ENTER])? y

It is now: 07-30-1990 at 04:28:41
You have been on-line for 7 minutes, 57 seconds
DON, Thanks and please call again!
```

APPENDIX E

Recommended Reading

The following is recommended reading for those who want more detailed information on telecommunications. If you would like a complete EcoLinking bibliography, you can download one from the MNS Online BBS at 518/381-4430.

Adams, Dennis and Mary Hamm. "Telecommunications and the Building of Knowledge Networks: Here Today, Much More Tomorrow." *Educational Technology,* Vol. 28, No. 9, September 1988, page 51.

Anonymous. *BITNET Overview*. Corporation for Research and Educational Networking, BITNET Network Information Center. July 13, 1989. 17 pages.

Frey, Donnalyn and Rick Adams. *A Directory of Electronic Mail: Addressing and Networks*, 2nd ed. Sebastopol, CA: O'Reilly and Associates, 1990. 420 pages.

Harper, Michael. "The Network Nation." *PC World*, Vol. 5, No. 8, August 1987, page 296.

Horvitz, Robert. "The Usenet Underground." *Whole Earth Review,* Winter 1989, page 112.

Jones, Paul. *What Is the Internet?* Academic Computing Services, University of North Carolina, Chapel Hill, NC. February 8, 1990. Six pages.

Kovacs, Diane. *The Directories of Academic E-Mail Conferences.* Available on BITNET or the Internet. 1991. About 100 pages.

Kroll, E. *The Hitchhiker's Guide to the Internet.* University of Illinois, Urbana. 1118. September 1989.

LaQuey, Tracy L. *The User's Directory of Computer Networks.* Bedford, MA: Digital Press, 1990. 630 pages.

Malkin, G., et al. *FYI on Questions and Answers to Commonly Asked "New Internet User" Questions.* NSF Network Service Center. September 7, 1990. 24 pages.

O'Reilly, Tim and Grace Todino. *Managing UUCP and Usenet.* Sebastopol, CA: O'Reilly and Associates, 1989. 269 pages

Quarterman, John. *The Matrix: Computer Networks and Conferencing Systems Worldwide.* Bedford, MA: Digital Press, 1990. 720 pages.

Strangelove, Michael. *Directory of Electronic Journals and Newsletters.* Office of Scientific and Academic Publishing, Association of Research Libraries, 1527 New Hampshire Ave., NW, Washington, DC 1991. About 100 pages.

Wood, Lamont and Dana Blankenhorn. "State of the BBS Nation: Behold the Lowly Bulletin Boards, Now Encompassing the Globe." *BYTE,* Vol. 15, No. 1, January 1, 1990, page 298.

Index

A

Abbreviations, 26
Abercrombie, David, 222–23
Academic American Encyclopedia, 225, 264
Academic Index, 225
Access, America Online, fee for, 146
Acid rain, 188
Address, of BBS on FidoNet, 36
Agence France Presse, 272
Agencia EFE, 272, 288
Agribusiness USA, 226
Agricola, 225
Agriculture, 225, 227, 295–96
Agriculture Library Forum, 123
Agriculture/plant science, 123–24
Agrochemicals Handbook, 226
Air and climate, 188
Air quality, 125
Air/Water Pollution Report, 285
Alerts, 188
Alliance for Environmental Education, 194
Alternative Farming Systems Literature database, 124
AlterNet, 317
Alternex (Brazil), 177
Amateur Scientist BBS, 143–44
American Economic Association, 262
American Foundation for Biological Sciences (AFBS), 126
American Geological Institute, 263
American Men and Women of Science, 240–41
American Physical Society Bulletin, 138
American Society of Hospital Pharmacists, 227
American Water Works Association, 137
America Online, 5, 146, 268
    clipping service on, 275–81
    cost of, 148
    departments of, 148
    environmental resources of, 150
    getting online, 148
    navigating through, 149
    services of, 147
Amiga, 291
Analytical Abstracts, 241
Anderson, Owen T., 314
Animal behavior, 303–04
Animal rights, 73–74, 296
    in farming, 200
Announcements, 188
AP, 159, 268
AP Online, 269
Apple Access II, 291
Apple Computer, 7
Apple II, 291
Applied Genetics News, 282
Aqualine, 241
Archiving, 19–20
Area Business Databank, 235
ARPANET, 79
Arris, Lelani, 72–73
Articles, 63
Asbestos Control Report, 286
ASCII Express, 291, 292, 293
ASCII Pro, 291, 292, 293
ASCII set, 11
Associated Press, 159, 268
Associated Press Online, 269
Association for Progressive Communications, 178
Astronomy/space, 125–26
Asynchronous telecommunication, 10
Atari, 292
AT command set, 13
Atomic Energy Commission, 246
Automatic callback feature, 108
Availability, of Knowledge Index, 224

B

BackComm, 293
Baldeck, Charles, 175
Bashin, Bryan, 271–72
Batch transfer, 18
Bauchau, Vincent, 305
Baud rate, 13
BBS
    connecting to, 102
    registering on, 104
    validation survey on, 107–11
    welcome screen of, 102–04
Bean Bag, The, 128
Beebox, 74
Beekeeping, 74

343

BEE-L, 297
Behavioral ecology, 303–04
Beyond war, 189
Bibliographic database companies, online, 219–20
Bibliographies, collections of, 259
Biographies, 234
Bio Info conference, 212, 218
Biological Abstracts, 128, 260
Biological sciences, 126–28, 297–307
Biology, 68–69, 227
Biomechanics, 300
Bionet, 68–69
Bioprocessing Technology, 282
BioScience, 126
BIOSIS, 128, 260
BIO-SOFTWARE, 298
BIOSPH-L, 300
Biostatistics, 303
Biotech Business, 282
Biotechnology, 282–83, 301
BioTron, The, 126
Birds of America, 261
BITNET, 5, 28, 302
    cost of, 52
    getting online, 52
    mailing lists on, 57–62
Bits, 11
Bits per second (bps), 13
Black Bag BBS, 138
Bogert, Anton J. van den, 300
Bonine, John, 190–92
Books in Print, 226
Botany, 128–29
Braat, Arie, 302
Brittanica Software, 158
BRS/After Dark, 220, 236–39
BRS Colleague and Search Service, 236
BT Tymnet, 249
Buffalo Museum of Science, 128

Bulletin board list, 7
Bulletin boards
    connecting to, 100–101
    databases on, 120–22
    features of, 112–22
    file libraries on, 114–20
    finding, 101
    list of environmental, 122–44
    logging onto, 101–02
Bulletin board services (BBSs), 5, 99
BusinessWire, 272, 288
Bytes, 11

C

CAB Abstracts, 227
Cables, 14
Calendars, 189
California forests, 196
Cambridge Scientific Abstracts Life Sciences, 238
Campus Earth, 194
Carbon Copy Plus, 293
Cargill, John, 307
Cavers, 74
CCITT, 13
CD-ROM
    environmental titles on, 260–66
    how it works, 259–60
CD-ROM readers, 257
Center for Conservation Biology at Stanford, 178
Center for Exposure Assessment Modeling (CEAM), 140–41
CERFnet, 317–18
Chemical Engineering and Biotechnology Abstracts, 241–42
Chemical Information System, 248
Chemical Manufacturers Association, 248

Chemical properties database, 232
Chemical Referral Center, 248
Chemicals, 283–84
Chemical Safety NewsBase, 242
Chemistry, 74–75, 312
    analytical, 241
Chem-talk, 74–75
Christian Science Monitor, 235
ClariNet, 70
Clede, Bill, 174
Climate, 129
Climatedata NCDC 15 Minute Precipitation, 261
Climbing, 75
Clintonia, 128
Clipping service, personal electronic, 267
Commercial online services, 5
Commodore, 294
Commonwealth Agricultural Bureau, 227
Communications software, for America Online, 147
Communications surcharges, 164
CompendexPlus, 227
Comprehensive Core Medical Library, 238
Compton's Encyclopedia, 157–58
Compu-Farm BBS, 123
CompuServe, 5, 146
    clipping service on, 268–71
    conference areas on, 170
    cost of, 164
    databases on, 175–76
    getting online, 164
    member directory on, 170–71

Index   345

message boards on, 169–70
organization of, 164–65
services of, 163
Computer, 9–10, 145
Comtex, 288
Conference areas, on CompuServe, 170
Conference hall, on America Online, 151
Conferences
    on EcoNet, 185–200
    list of, 207–12 table
    on The WELL, 206–07
Conferencing, on The WELL, 213
ConflictNet, 177
Conservation, 130
Conservation Digest Newsletter, 198
Conservation Directory, 194
Consumer Drug Digest, 227
Consumer Drug Information Fulltext, 227
Contamination, toxic, 199
Controversial, 44
Coordinated list of chemicals (CLC) database, 140
Corporation for Research and Educational Networking (CREN), 51
COSMIC, 94
Cost
    of America Online, 148
    of BITNET, 52
    of CompuServe, 164
    of EcoNet, 179
    of FidoNet, 32–33
    of GEnie, 202
    of Internet, 81
    of Knowledge Index, 224

    of Usenet, 65
    of The WELL, 206
CPM machines, 294
Crosstalk, 293
Current Awareness in Biological Sciences, 238, 242
Current Biotechnology Abstracts, 230
Current Contents, 239
Current Contents: Agriculture, Biology, and Environmental Sciences, 238
Current Contents: Life Sciences, 239
Current Contents on Diskette, 249
Cyclopean Gateway Service, 249

D
DASnet, 315–16
Databases
    bibliographic, 5
    on bulletin boards, 120–22
    on CompuServe, 175–76
    of Knowledge Index, 224–36
    online scientific on Internet, 93–94
    specialized, 247–56
Databits, 101
DataPac, 22
Data ports, 10–11
Deep Sea Drilling Project, 261
Defense Data Network Network Information Center, 82
Defense Issues, 44–45
Department of Commerce, 239
Department of Defense, 79

Department of Physics and Astronomy, 139
Departments, of America Online, 148
Deutsch Press—Agentur, 272
Development, 189
DIAL n' CERF, 317–19
Dialog Information Services, 221
Dictionary of Organic Compounds, 232
Dictionary of Organometallic Compounds, 232
Directories of Academic E-Mail Conferences, 88
Directory of Electronic Journals and Newsletters, 86
Discovery Environmental Data, 261
Dissertation Abstracts Online, 230
Dodell, David, 37, 138
Domain addressing, 82
Downloading, 18
    file on BBS, 116–20
Duplex, 17

E
EARN, 51
Earth Echo, The, 42
Earthlife, 186
Earthquake-tracking network, 152–53
East-west, 189
Easyplex, 175
Echoes, 32
    environmental, 38–49
    environment and science, 41–42
    geography, 43
    health and safety, 43–44
    list of Fido, 40–49
    news, 45–47
    politics, 44–45

recreation, 48
technology, 48–49
Echomail, 32
Ecological modeling, 130
EcoNet, 5, 64, 146, 177
    commercial networks available through, 184–85 table
    conferences on, 185–200
    cost of, 179
    environmental resources of, 180
    getting around, 179
    getting online, 178
    organizations you can reach through, 181–83 table
EconLit, 262
Economic Literature Index, 231
Economics, 262
Eco System BBS, 131
Education, 194
Education and research, 192
Electric Power Industry Abstracts, 243
Electronic mail, 160–61, 212, 218
    on CompuServe, 175
    on GEnie, 204
    through gateways, 183–85, 315–18
Electronic newsletter distribution, 72–73
Electronic Whole Earth Catalog, 262
E-mail, 28
    on BITNET, 54–56
    on Internet, 82–83
Emergency Management Information Center, 250
Emergency Management Institute, 250
Energy, 192–93, 284–85, 307

Energy Conservation News, 284
Energy Daily, The, 284
Energy Library, 246
ENERGYLINE, 243
Energy Report, The, 284
Energy Research and Development Administration, 246
Engineering Literature Index, 231
Enviroline, 243
EnvironChat conference room, 157 *fig*
EnviroNet, 130
Environment, 131, 194–96, 285–87, 312–13
Environmental behavior, 302
Environmental Bibliography, 262
Environmental Forum, The, 150–54
Environmental Grantmakers Association Directory, 195
Environmentalism, definition of, 3
Environmental Issues, 41
Environmental Law Alliance Worldwide (E-LAW), 190–92
Environmental networks and associations, 180–83
Environmental News Weekly Reader, 156, 275
Environmental publications, on NewsNet, 282–88
Environmental research, bulletin boards for, 122–44
Environmental Research Foundation (CERF), 135
Environmental resources of America Online, 150
    on CompuServe, 167
    of EcoNet, 180

    of GEnie, 203
    of The WELL, 212
Environmental titles on CD-ROM, 260–66
Environment and science, echoes on, 41–42
Environment conference, 212, 213–17
Environment News Service, 268, 289–90
Environment Week, 286
EPA, 125, 140, 141
Epidemiology, 303
Error-checking protocols, 18
Eschallier, Phil, 64
Ethology, 75
Etiquette, online, 24–26
European Bank of Computer Programs on Biotechnology, 301-02
European Molecular Biology Laboratory, 251
Even parity, 17
Evolution, theory of, 304
Evolutionary Mechanism Theory Discussion, 42
Excerpta Medica Library Service, 263
Executive News Service, 269
Extension abbreviation, 20

F
Facts on File News Digest, 263
Farm Business Management Branch, 123
Farming, animal rights in, 200
Federal Energy Administration, 246
Federal Power Commission, 246
Federal Register Abstracts, 239
FELINE-L, 305–06
FEMA, 132, 250

Feminist approach, 196
FidoNet, 5, 28, 64
    cost of, 32–33
    gateways to other networks on, 37–38
    getting online, 32
    organization of, 33–36
File libraries
    on bulletin boards, 114–20
    on CompuServe, 168–69
File transfer, 17–19, 32
    on Internet, 83–84
File-transfer protocol, 169
Firedoc, 250–51
FireNet Leader, 133
Flaming, 40
Flash, 292
Flora Online, 128
Foo, Eng-leong, 301
Food and agriculture, 196
Food Science and Technology Abstracts, 232
Forest, 243
Forests, 196–97
Forum, The, 126
Forum, organization of CompuServe, 168–71
Framework Convention on Climate Change, 72
FredsNaetet (PeaceNet Sweden), 177
Friends of the Earth, 178
Ftp, 83–84
Full duplex, 17

G

GaiaNet, 27
Galaxy Star Link, 22
GenBank, 94, 251
General Electric Company, 201
Genetic engineering, 195
Genetics Computer Group software, 306
Genetic Technology News, 283
GEnie, 5, 146, 268
    clipping service on, 272–75
    cost of, 202
    electronic mail on, 204
    environmental resources of, 203
    getting around, 202–03
    getting online, 201–02
GenPept, 251
Geography, 43, 313
Geology, 132
GeoMechanics Abstracts, 244
GeoRef, 244, 263
Germplasm Resources Information Network, 252
Global networks, 5, 27–29
Global warming, 188
GO commands, in CompuServe, 165
Good Earth Forum, 173
Government agencies, 195
Government Reports Announcements and Index, 264
GPO Publications Reference File, 232
Greenhouse Effect Report, 286
Green movement, 197
GreenNet (England), 177
Green Party, 195, 197
Greenpeace, 130, 178, 195, 198

H

Half duplex, 17
Harlan, John B., 310
Harper, Rob, 298
Hayes-compatible, 13
Hayes Microcomputer, 7
Hayes Terminal Program Micromodem II, 292
Hazardous materials, 250
Hazardous Materials Information Exchange, 132–33
Hazardous Substance Databank, 252
Hazardous waste, 132–36, 228
Hazardous waste incinerator, 193, 199
Hazardous Waste Management, 41
Hazardous Waste News, 283
Health, 197
Health and safety, echoes on, 43–44
Health Physics Society Conference, 42
Heath/Zenith, 292
Heilbron, 232
Help, getting, 6–8
Herpetology, 75–76, 136
HerpNet, 136
HICN News, 138
Hiking, Mountain Climbing, and Camping, 48
HomeoNet, 177
House of Representatives committees, 195
Hubs, 36
Hunger Conference, 43
Hydrology, 137
HydroWire, 285
HyperACCESS, 293

I

IBM, 293
IBM PC, CD-ROM on, 259
Index to Dental Literature, 234
Index of Economic Articles, 262
Index Medicus, 234

Indian Affairs, 41
Indigenous peoples, 198
Indoor Air Quality Update, 286
Industrial Health and Hazards Update, 288
Inet/ddn, 70–71
Information Manager, CompuServe, 166
Information Power, 48–49
Institute for Global Communication, 177, 178
Institute for Molecular Biology, 94
Institute for Scientific Information, 249
Integrated Pest Management BBS, 124
Integrated Risk Information System (IRIS), 252, 253
IntelliGenectics, 251
Intelliterm, 293
International Forum, 42
International Nursing Index, 234
International Union for the Conservation of Nature, 178
Internet, 5, 29
    cost of, 81
    via FidoNet, 37–38
    getting online, 80–81
    resources of, 87
Internet Resource Guide, 82
Invest/Net, 288
IRIS list of toxicity numbers, 223
Israel, energy research in, 307
ISTP Search, 244
IuBio Archive for Molecular and General Biology, 94

J
James, Robert C., 303
Jennings, Tom, 31
Job listings, 189
Jones, Paul, 313
Journal of Economic Literature, 262
Journal of Student Research, 143
JRComm, 291

K
Kermit, 18, 295
Killifish, 76
Kilobyte, 11
Knowledge Index, 220, 221–36
    availability of, 224
    cost of, 224
    databases of, 224–36
Knutsen, Andrew, 309
Kodak, 280–81
Kristofferson, Dave, 299
Kyodo News Service, 272, 288

L
Labor and Union News Conference, 43
Law, 288
Legal Resources Index, 233
Lewis, Sanford, 186
Libraries, online on Internet, 91–93
Life Sciences Collection, 233
LiMB Database, 94
Line, Les, 174
Listing of Molecular Biology database, 94
List servers, on BITNET, 56–62
LiveWire NewsRoom, 273
Local National Regional News, 45
Lockheed Corporation, 221
Logging, 191
Longitude and latitude, tracking, 152–53
Los Angeles Times, 235
Lync Communications Software, 293

M
Macintosh, 293–94
    CD-ROM on, 259
MacKnowledge, 294
Macros, 16
MacTerminal, 296
Magazine Index, 234
Mail
    on bulletin board, 113–14
    on Internet, 82–83
Mailing lists, 51, 88–89
    on BITNET, 57–62
    on UUCP, 73–78
Main menu, of bulletin board, 112
Mandile, Tony, 174
Mapping, 43
Marine mammal captivity, 198
Marquis Who's Who, 234
Marshall Space Flight Center, 125
Maxwell Online, 239
McClain, Dennis, 39–40
McGraw-Hill, 268
Mcterm, 292
Media, 197–98
Media Transcription Service, 46
Medical Round Table, 203–04
Medicine, 137–38
MEDLARS, 252
MEDLINE, 234, 255, 264
Member directory on CompuServe, 170–71
Menus, on CompuServe, 165
Message boards, 153–54
    on America Online, 151
    on CompuServe, 169–70

Michigan Water Pollution Control Association, 137
MicroPhone, 296
MILNET, 79
MNS Online BBS, 7, 131
Modcom, 292
Modems
    communicating with, 10–14
    duplex, 12
    smart, 13
ModemWorks, 294
Modulation/demodulation, 12
Molecular biologists, 251
Molecular biology, computer-aided, 306
Monitor display, setting up on BBS, 105–07
Monteiro, Tony, 303
Moor, Aldo de, 53–54
Mossberg, A.E., 313
Mouse, 149
Mousewrite, 292
Mueller, Scott Hazen, 310
Multinational Environmental Outlook, 287

N
NASA, 94
NASA SpaceLink, 125
National Agricultural Library, 123, 225
National Biological Impact Assessment Program BBS, 127
National Birding Line Cooperative, 139
National Directory, The, 264
National Ecology Research Center, 130
National Emergency Training Center Learning Resource Center, 250
National Environmental Data Referral Service, 239

National Fire Academy, 250
National Geophysical Data Center, 261
National Geographic Kids Network, 143
National Germplasm Resources Laboratory, 252
National Library of Medicine, 222, 234, 252, 255, 264
National Newspaper Index, 235
National Oceanographic and Atmospheric Administration (NOAA), 129
National Plant Germplasm System, 252
National Science Foundation, 79, 127
National Science Teachers Association, 144
National Space Society BBS, 126
National Technical Information Service, 245, 264–65
National Toxics Campaign Fund, 186
National Wildlife Federation's Conservation Directory, 178
Native American Controversy, 47
Native American issues, 198
Native American NewsMagazine, 47
NativeNet, 76
Natural history, 297–307
Natural Products ALERT, 253–54
Nature, 138–39
Navigating, through America Online, 149
Needham, Rick, 172
Netmail, 32

NetNorth, 51
Nets, 36
Network Earth, 171–72
Networking, equipment necessary for, 4
Networking for Change, 45
Networks, global, 5, 27–29
New Grolier Electronic Encyclopedia, 264
Newlists, 76
News, echoes on, 45–47
Newsbytes, 275, 277
NewsCom, 76–77
Newsearch, 235
Newsgroups, 29, 63, 65–67
    alternative, 68–78
Newsletters, 198
NewsNet, Inc., 268, 277–80
    environmental publications on, 282–88
News of the U.S. and World, 46
Newspaper Library, 270
News services, online, 220
NewsWatch, 159, 276
News wire services, 159
New York Times, 235
Nicarao (Nicaragua), 177
Nitrogen fixation, biological, 303
Nixpub, 65
Node, 36
Nonpoint Source Information Exchange BBS, 142
NSFNET, 79, 82
NTIS, 245, 264–65
Nuclear fusion, 310
Nuclear incidents, 250
Nuclear industry, in Europe, 193
Nuclear weapons production facilities, 199

O
Occupational safety, 174–75

Ocean policy, 77
Odd parity, 17
Office of Solid Waste and
    Emergency Response, 133
Offshore Gas Report, 285
Oil Spill Intelligence Report,
    287
Online
    on America Online,
        148
    on BITNET, 52
    on CompuServe, 164
    on EcoNet, 178
    on FidoNet, 32
    on GEnie, 201–02
    getting, 5
    on Internet, 80–81
    on Usenet, 64–65
    on The WELL, 205
Optical Publishing Association, 257
ORBIT, 220, 239–47
Organization, of CompuServe, 164–65
Ornithology, 139
Orzech, Mary Jo, 299
OSHA Computerized Information System, 254
Osprey's Nest, 139
OSWER BBS, 133–34
Outdoor Association of
    America, 174
Outdoor Forum, 173–75

P

Packet switching networks, 21
Paper, Printing, Packaging
    and Nonwovens
    Abstracts, 245
Paper Chase, 255
Parity, 16–17, 101
Parks, 197
Password, 104
Patnaik, Deba, 303
PC Pursuit, 22

PC-Talk III, 293
PeaceNet, 177
Pegasus (Australia), 178
Penn State's College of Agriculture, 94
PENpages, 94
PESTDOC, 245
Pesticides, 132–36, 196, 247
Peyton, Dave, 173
Phillips, Dave, 302
Physics, 139–40, 307–10
Place Name Index, 265
Plant genetics resources, 252
Plants, 265
Politics, 45
    echoes on, 44–45
Pollution, 188
Pollution abatement, 140–43
Pollution Abstracts, 236
Pollution Prevention Information Clearinghouse 141–42
Pollution Prevention Information Exchange System, 141–42
Polymer physics, 309
Population, 189, 198
Population Biology, 304–05
Power, 246
Pozar, Tim, 37
PR Newswire, 273
Procomm, 293
Program for Collaborative
    Research in the Pharmaceutical Sciences, 253
Publications, 225
Public conferences, 156–57
Public health, 137–38
Public message boards, 114

Q

Q-Modem, 293

R

RACHEL, 135–36

Rainforest Action Groups,
    197
Rain Forest Network Bulletin, 53–54
Rain forests, 197
Radiation Safety, 44
Readers, 65
Read-only format, 258
Recreation, echoes on, 48
Recycling, 192, 199
    fraud in, 228–29
Regions, 33–36
Registering, on BBS, 104
Registry of Toxic Effects of
    Chemical Substances,
    252, 253
Remote Access Chemical
    Hazards Electronic
    Library, 135
Researchers, locating, 94–96
Reuter's, 159, 268
Reynolds, Joe, 174
Rizzo, John, 259
Roberson, Chip, 297
Robertson, Tim, 191
Royal Society of Chemistry,
    230
Rutgers Cooperative Extension BBS, 124

S

Saarikko, Jarmo, 346
SafetyNet Forum, 175
Safety Professional's Forum,
    44
Safety Science Abstracts, 246
St. Joseph's Hospital BBS, 137
Sandberg, Steffan, 172
Scheutjens, Jan, 309
Science conference, 212, 217
Science education, 143–44
Science Line BBS, 144
*Science Line Navigation
    Guide,* 144
Science magazine, 238

Science/Math Educational Forum, 172–73
SCIENCE net, 255–56
Scripts, 16
Seas and waters, 198
SEEDS Project, 39–40
Serial ports, 10–11
Services, online, 146
SFER-1, 77
Shareware, 20
Shein, Barry, 314
Sierra Club, 178
Sierra Club National News Report, 198
Skyland, 138–39
Smartcom, 294
Smart modems, 13
Smileys, 26
Smith, Dick, 308
SMTP, 82
Softerm 2, 292
Software
    communications, 14–16
    related to biological sciences, 298
Software library, 154–56
    on America Online, 151
Solstice magazine, 196
South Florida Environmental Reader, 313
Southwest Research Data Display and Analyses System, 94
Soviet activists, 189, 195
SovNet-L, 309
Space-activists, 77
Space Environment Laboratory, 129
Spacemet Central/Physics Forum, 139–40
Space science, 313–14
Spelunker's Forum, 48
Spelunking, 74

Sprafka, Sandy, 296
SprintNet, 21
Standard Pesticide File, 247
Start bits, 16
State Regulation Report Toxics, 283
Stop bits, 16, 101
Student Environmental Action Council, 312
Supercomputers, 89–90
Superfund, 287
Support Center for Regulatory Air Models BBS, 125
Sustainable Agriculture, 41
SWISS-PROT, 251
SysOps, 32, 111

T
Tandy, 296
TAXACOM, 128–29
Taxonomic Reference File, 128
Technology, 199
    echoes on, 48–49
Technology Education, 49
Technology issues, 311
Telecommunications, basics of, 5
Telephone, saving on costs, 21–22
Telix, 293
Telnet, 85–87
Term-Talk, 292
Termworks, 294
Time limitation, on BBS, 111–12
Timetable of History: Science and Innovation, 265
Toxic chemicals, 197
Toxic cleanup, 199
Toxic Release Inventory, 222, 252, 253
Toxics, 132–36, 199
Toxics Materials News, 284
TOXNET, 252

Toxnet computer system, 222
Tropical Agriculture, 247
Turner Broadcasting System, 171
Tymnet, 21

U
UFGate, 37
UltraTerm, 294
Union of Concerned Scientists, 196
United Press International, 159, 268, 273
U.S. Fish and Wildlife Service National Ecology Research Center, 130
United States Geological Survey BBS, 132
United States-South Africa, 46–47
University of Wisconsin Extension, 124
U.N.-NGOs, 46
UPI, 159, 268, 288–89
Uploading, 18
Urban Phytonarian, 265
U.S.A.-EuropeLink, 47
Usenet, 5, 29, 212, 218
    cost of, 65
    getting online, 64–65
User group, 6
User manual, for The WELL, 207
US Sprint, 22
USSR, 311
UUCP, 63, 71–78
UUNet, 64, 316–17

V
Validation survey, on BBS, 107–11
Valles, Asuncion, 312
VersaTrem, 294
Virginia Polytechnic Institute and State University, 127

Virus checking, 20–21
Vitis, 232
Vulcan's Computer Monthly, 6, 7, 101

W

Waiver, 108
Wall Street Journal, 235
Washington Post, 235, 268
Waste and Wastewater Network, 137
Wastewater technology, 241
Water Research Centre, 241
Water resources, 137
Web, The (Canada), 178
Weinstein, Michael P., 204
Welcome screen, on BBS, 102–04
WELL, The, 5, 64, 146
    conferences on, 206–07, 213
    cost of, 206
    environmental resources of, 212
    getting around, 207–12
    getting online, 205
Wener, Richard, 303
White Knight, 294
White Pages, 95
Whole Earth Catalog, 205, 262
Whole Earth 'Lectronic Link, 205
Wilderness, 200
Wilderness Experience, 48
Wildlife, 200
Wildnet, 78
Wind energy, 192
Windstar Connections, 193
Woltring, Herman J., 302
Wood products industry, 243
Woods, Wendy, 275
World Factbook, 266
World hunger, 189
World Rainforest Report, 197
World Resources 1990–91, 261

X

Xinhua News Agency, 273, 288
Xmodem, 18

Y

Ymodem, 18

Z

ZLYNK,II, 292
Zmodem, 18
Zones, 33
Zoological Records, 128
Zterm, 296
Zwaren de Zwarenstein, Joseph van, 307, 308

Want to Stay Ecolinked?

Become part of the *EcoLinking* community free!

Just send us your name and address and you will have free access to the MNS ONLINE BBS *EcoLinking* section (518/381-4430), as well as be the first to learn about updates, new services, and future editions of *EcoLinking*.

I am interested in the following:

- ❏ Networks
- ❏ Bulletin boards
- ❏ Online services
- ❏ Databases
- ❏ Bibliographical services
- ❏ CD-ROM
- ❏ All of the above

Also, please pass along comments about this edition of *EcoLinking*. We are always looking for new and exciting stories and new services. You can purchase bulk copies of *EcoLinking* at a discount for your organization by contacting Paula Baker at Peachpit Press (800/283-9444).

Name ────────────────────────────

Address ──────────────────────────

City, State, Zip ──────────────────

Network or online ID ──────────────

Where did you purchase *EcoLinking*? ─────

Cut & Send To:
EcoLinking
PO Box 463
Schenectady, NY 12301-0463

More from Peachpit Press...

Canned Art: Clip Art for the Macintosh, 2nd Edition
Erfert Fenton and Christine Morrissett
Voted Best Book of 1990 by *MacWeek*'s Ezra Shapiro, this second edition is an encyclopedia of over 15,000 images, meticulously indexed to help the graphics professional make qualified purchases. A special section covers clip art management, file formats, and printing options, and provides tips on cutting, pasting, and modifying images. The book contains tear-out coupons worth over $1,000 in discounts on clip art packages. *(640 pages)*

Desktop Publisher's Survival Kit
David Blatner
This book/disk package provides insights into publishing on the Macintosh: troubleshooting print jobs, working with color, scanning, and selecting fonts. It also covers everything from graphics file formats and digital fonts to word processing, color, typography, style sheets and printing techniques. A disk containing 12 top desktop publishing utilities, 400K of free clip art, and two fonts is included. *(184 pages)*

Desktop Publishing Secrets
Robert Eckhardt, Bob Weibel and Ted Nace
This compilation offers updated and expanded versions of the 500 best tips and Q&A answers from *Publish* magazine. *The Publish Book of Tips* covers all the major publishing programs on both the PC and Macintosh platforms and includes *Publish*'s most experienced writers. *(536 pages)*

Help! The Art of Computer Technical Support
Ralph Wilson
This book overviews the technical support industry, analyzes the dialog between the tech support worker and the user in trouble, explains how to set up and manage a support operation, and provides profiles of successful technical support operations. *(232 pages)*

The Little DOS 5 Book
Kay Nelson
All you need to know about DOS 5, organized concisely and written in plain English. The book is packed with plenty of tips as well as an easy-to-use section on DOS commands that explains things with everyday, practical examples. Also covers DOS basics, working with files and directories, disk management, and more. *(160 pages)*

The Little Mac Book, 2nd Edition
Robin Williams
Praised by scores of magazines and user group newsletters, this concise, beautifully written book covers the basics of Macintosh operation. It provides useful reference information, including charts of typefaces, special characters, keyboard shortcuts, and a special update on System 7. *(184 pages)*

The Little Mac Word Book
Helmut Kobler
For users new to Microsoft Word or for experienced users who want to familiarize themselves with the features of version 5.0, this book is just the ticket. In addition to discussing Word basics, it provides concise and clear information about formatting text; using Word with Apple's new System 7 operating system; taking advantage of Word's writing tools, including its spelling checker, thesaurus and grammar checker; setting up complex tables, and much more! *(240 pages)*

The Little QuicKeys Book
Steve Roth and Don Sellers
A handy guide to CE Software's QuicKeys 2.0, this book explores the QuicKeys keysets and the different libraries QuicKeys creates for each application; shows how to link together functions and extensions; and provides an abundance of useful macros. *(288 pages)*

The Little System 7 Book
Kay Yarborough Nelson
In clear, simple English, this first-rate reference book lays out everything you need to know to take advantage of System 7's virtual memory, desk accessories, and the new finder and Control Panel. It covers TrueType, tricks for multitasking, and ways to customize your system. *(160 pages)*

The Little WordPerfect Book
Skye Lininger
Teach yourself the basics of WordPerfect 5.1 in less than an hour. This book shows you just enough to start creating simple letters, memos, and short reports—fast. Gives easy-to-understand, step-by-step instructions for setting page margins, typing text, navigating with the cursor keys, basic editing, printing, and online help. *(160 pages)*

The Little WordPerfect for Windows Book
Kay Nelson
Ideal for any WordPerfect for Windows user, this book gives you the basic skills you need to create simple documents and get familiar with WordPerfect's new Windows interface. Also covers more advanced topics, such as formatting pages, working with blocks of text, using different fonts, and special features including WordPerfect's new mail merge, tables, equations, indexes, and footnotes. *(200 pages)*

The Mac is not a typewriter
Robin Williams
This bestselling, elegant guide to typesetting on the Mac has received rave reviews for its clearly presented information, friendly tone, and easy access. Twenty quick and easy chapters cover what you need to know to make your documents look clean and professional: em dashes, curly quotes, spaces and indents, special characters, hyphenating line breaks, and more. *(72 pages)*

The PC is not a typewriter
Robin Williams
PC users can now learn Robin Williams' secrets for creating beautiful, professional-looking type. In less than 100 pages, this book explains both the principles and the logic behind the techniques developed for professional typesetting. Covers punctuation, leading, special characters, kerning, fonts, and more. *(96 pages)*

PageMaker 4: An Easy Desk Reference (Mac Edition)
Robin Williams
Useful for both beginners and advanced users, this book uses a unique, three-column format to answer any PageMaker question as quickly as possible. *(784 pages)*

Safe Computing: Understanding and Avoiding Computer Health Hazards
Rebecca Rosen Lum
Your computer may be hazardous to your health. As the number of computer-related ailments has increased, so has public concern about computers and safety. *Safe Computing* is an invaluable resource for companies seeking informed choices about workplace design and job structuring and for computer users who wish to protect themselves from injury. Topics include vision problems, back and neck injuries, electromagnetic radiation and repetitive strain injuries. *(300 pages, Spring 1992)*

Winning! The Awesome and Amazing Book of Windows Tips, Traps, and Sneaky Tricks
John Hedtke
Now you can have even *more* fun with Windows. Here's the inside story on games like Tetris, Minesweeper, Solitaire and Reversi. This book includes the complete rules and instructions for each game, as well as hidden features of many games. Finally, it shows some sneaky tricks learned directly from the games' programmers that let you rack up huge scores in record time. *(232 pages)*

Order Form
(800) 283-9444 or (510) 548-4393
Fax: (510) 548-5991

#	Title		Price	Total
	Canned Art: Clip Art for the Macintosh, 2nd edition	MAC	29.95	
	Desktop Publisher's Survival Kit (w/disk)	MAC	22.95	
	Desktop Publishing Secrets	PC MAC	27.95	
	EcoLinking	PC MAC	18.95	
	HELP! The Art of Computer Technical Support	PC MAC	19.95	
	The Little DOS 5 Book	PC	12.95	
	The Little Mac Book, 2nd edition	MAC	14.95	
	The Little Mac Word Book	MAC	15.95	
	The Little QuicKeys Book	MAC	18.95	
	The Little System 7 Book	MAC	12.95	
	The Little WordPerfect Book	PC	12.95	
	The Little WordPerfect for Windows Book	PC	12.95	
	The Mac is not a typewriter	MAC	9.95	
	The PC is not a typewriter	PC	9.95	
	PageMaker 4: An Easy Desk Reference	MAC	29.95	
	Safe Computing (due in Spring 1992)	PC MAC	14.95	
	Winning! The Awesome and Amazing Book of Windows Game Tips, Traps, & Sneaky Tricks	PC	14.95	

Tax of 8.25% applies to California residents only. UPS ground shipping: $4 for first item, $1 each additional. UPS 2nd day air: $7 for first item, $2 each additional. Air mail to Canada: $6 for first item, $4 each additional. Air mail overseas: $14 each item.	Subtotal
	8.25% tax (CA only)
	Shipping
	TOTAL

Name			
Company			
Address			
City		State	Zip
Phone		Fax	
❏ Check enclosed	❏ Visa		❏ MasterCard
Company purchase order #			
Credit card #		Expiration Date	

Peachpit Press, Inc. • 2414 Sixth Street • Berkeley, CA • 94710
Your satisfaction is guaranteed or your money will be cheerfully refunded!